QUIPS, QUOTES AND QUANTA

2nd Edition

*AN
ANECDOTAL
HISTORY
OF PHYSICS*

D0869465

QUIPS, QUOTES AND QUANTA

2nd Edition

AN ANECDOTAL HISTORY OF PHYSICS

ANTON Z CAPRI

UNIVERSITY OF ALBERTA, CANADA

 World Scientific

NEW JERSEY • LONDON • SINGAPORE • BEIJING • SHANGHAI • HONG KONG • TAIPEI • CHENNAI

Published by

World Scientific Publishing Co. Pte. Ltd.

5 Toh Tuck Link, Singapore 596224

USA office: 27 Warren Street, Suite 401-402, Hackensack, NJ 07601

UK office: 57 Shelton Street, Covent Garden, London WC2H 9HE

Library of Congress Cataloging-in-Publication Data
Capri, Anton Z.
  Quips, quotes, and quanta : an anecdotal history of physics / by Anton Z. Capri. -- 2nd ed.
  p. cm.
  Includes bibliographical references and index.
  ISBN-13: 978-981-4343-47-3 (pbk. : alk. paper)
  ISBN-10: 981-4343-47-1 (pbk. : alk. paper)
  1. Physics--History. 2. Physics--Anecdotes. I. Title.
  QC7.C2354 2011
  530.09--dc22
                                2011001516

British Library Cataloguing-in-Publication Data
A catalogue record for this book is available from the British Library.

For photocopying of material in this volume, please pay a copying fee through the Copyright Clearance Center, Inc., 222 Rosewood Drive, Danvers, MA 01923, USA. In this case permission to photocopy is not required from the publisher.

Printed in Singapore by World Scientific Printers.

To all my colleagues who, over the years, have told me their favourite anecdotes and especially to all those colleagues who have patiently listened to me repeat these anecdotes.

Preface to the Second Edition

Since the appearance of the First Edition I have received numerous letters containing suggestions and further anecdotes from colleagues and readers around the world. I want to thank all of them for their contributions. Particularly gratifying was a note from Marlies Green, the wife of the late Professor Bert Green of the University of Adelaide, who had provided me with some Dirac anecdotes. Here is some of what she wrote.

"I too remember many of the names you mentioned in the book, as Bert often talked about them, but of course Dirac's story is still vividly in my mind, as I was present at the beach as well. We often had a good laugh about it."

Her note provides a clue as to part of the reason I wrote this book. The older generation—my generation—of physicists who still met some of the founders of quantum mechanics and twentieth century ideas in physics are dieing out. I believe it is important to keep these stories alive so that future generations of physicists can have some understanding of the personalities involved in creating the subject they love and pursue.

In writing this book I also wanted to bridge part of the gap between what C.P. Snow calls "The Two Cultures" and bring to those of the general public interested in such matters a picture of what modern physics is and how it came about. Ironically, what is called "Modern Physics" in university curricula is now almost a century old and in some cases—as in the theories of relativity—even more than a century old. I did not, however, want to write a scholarly book with detailed references that would distract the reader from the flow of the story and the personalities involved. Besides, I did not always have knowledge of the source of the story. I wanted the reader to see that these physicists and mathematicians, although most exceptional in their abilities, had interests and desires not much different from the rest of us. Many also had a humorous side and human frailties, even though they were exceptional in their abilities.

If parts of the book reads like a hagiography, I do not apologize. These creators of modern physics were truly extraordinary and I could not, nor tried to, hide my admiration for them.

A. Z. Capri

Edmonton, Alberta.

December, 2010.

www.antoncapri.com

Preface to the First Edition

In writing this book I have been helped by numerous colleagues who, knowing of my interest in anecdotes, have over the years passed many of them on to me. It is not possible to list all of these contributors since in many cases I do not even remember who told me what. There are however, a few individuals whose contributions were substantially more than just relating an anecdote to me.

Professor M. Razavy not only provided me with continuous encouragement, but also with numerous references to historical writings. His encyclopedic knowledge has been a great help.

Professor H.S. Green, one of Max Born's last students, provided me with several anecdotes of Born and Dirac. As far as I know these have not appeared anywhere else.

During the early stages of writing this book I received moral support from Mary Mahoney-Robson, editor of the University of Alberta Press. I wish to express my appreciation for her encouragement and advice.

Finally, I am most indebted to Professor Edward W. Pitcher of the English Department of the University of Alberta who so generously gave of his time to review a substantial portion of an early draft of the manuscript. His comments have been invaluable in making this book flow somewhat more smoothly.

The various translations into English are, unless stated otherwise, my own. Except with the poetry, I have tried to maintain literal translations.

A. Z. Capri

Edmonton, Alberta.

April, 2007.

Contents

Chapter 1

Prologue

"When telling a true story one should not let oneself be overly influenced by the haphazard occurrences of reality." von Kàrman

My first encounter with ongoing research and physics humor was in 1958 while I was a second year undergraduate at the University of Toronto. Arthur L. Shawlow (1921 – 1999), co-inventor of the maser had come to give a seminar on the newest of physics discoveries, the laser. The highlight was a demonstration of this device. After a fifteen-minute period of "pumping" the laser all our eyes were directed to one of the walls in the big lecture theatre of the McLennan Laboratory where, for a fraction of a second, a very bright red spot appeared. I was not impressed. The only thing that impressed me was the large rubies—about 12 cm long with a diameter of 2 cm—that Shawlow had grown in his lab to make the lasers. I also do not remember much of what he said because I understood almost nothing. Nevertheless, I did learn that the maser was invented first and that maser was an acronym for **m**icrowave **a**mplification by **s**timulated **e**mission of **r**adiation. However, as Dr. Shawlow quipped, it should really stand for "**M**oney **a**cquisition **s**cheme for **e**xpensive **r**esearch". This was the first time in my life I was exposed to a form of physics humor. After that I was on the lookout for more of this type of humor and have managed over the past forty-five years to collect a large number of quotes by and anecdotes about famous and not-quite-so-famous physicists.

This was not difficult to do because, wherever physicists gathered to chat or have a drink, sooner or later, some anecdote about one of the founders of their subject was sure to be retold. Scathing barbs sometimes were intended to slay the living. Funny tales raised the dead or brought

1

life to physicists merely mistaken for dead.

A most unforgettable experience, from my days as a graduate student at Princeton, was to meet Professor Eugene Wigner (1902 – 1995). His politeness was reminiscent of the bygone era of the Austro-Hungarian empire, of deep bows and the kissing of ladies' hands.

One of the less pleasant experiences as a graduate student at Princeton was the oral portion of the General Examination, a sort of inquisition into those areas of the student's ignorance that had been revealed during the written portion of the exams. However, to have Wigner on one's examining committee was a stroke of good luck. The reason for this was that, whenever another professor asked a question, Wigner would not let the candidate answer since he believed the question was too vague. Then Wigner would proceed to "clarify" the question. This clarification of the question continued until the answer became self-evident.

Before the construction of Jadwin Hall, physics at Princeton was housed in the old Palmer Lab, which was connected by a corridor to Fine Hall, the building that housed mathematics. At each end of this corridor was a set of doors. It was well known among the graduate students that Wigner always insisted on opening doors for other people. Thus, the biggest challenge anyone could face at Princeton was to get Wigner to precede him through both sets of doors connecting Fine Hall with Palmer. There was a standing bet among the graduate students that this could not be achieved. Many tried. None succeeded.

Another long-established tradition at Princeton was the afternoon tea. The teapots, coffee urns and cups were placed in the hall so that everyone might help himself and enter the main room, that is until Wigner arrived. This true gentleman of the old school would insist on serving everyone behind him before taking his turn.

At Princeton, as at most universities, parking was at a premium. Thus, only essential people such as janitors were allowed to park close to Palmer Laboratories. One morning Wigner was late for a lecture and attempted to park near Palmer to reach his lecture on time. Of course, the parking attendant would not let a mere professor and Nobel laureate park there and told Wigner so in no uncertain terms. Wigner tried desperately to explain that he was late for a lecture. To no avail. Again, somewhat more forcefully, he tried to explain that a whole classroom of students was waiting for him. The attendant remained firm. Finally in exasperation Wigner exclaimed, "Go to hell!" But his politeness returned and before rushing off to his lecture he turned and added, "Please."

On the other hand, during seminars when visiting physicists presented their latest results, Wigner's questions were loaded. Fortunately, in most cases, Valentine (Valya) Bargmann (1908 – 1989) who always sat beside Wigner in the front row, would answer the question before it reached the speaker. There were a few occasions when this did not happen and these were memorable. One such occasion commenced with Wigner's usual hesitant, polite beginning, "Excuse me please, may I ask a question?"

"Why yes, please."

At this point Wigner stood up and pointing at an equation on the board asked, "Ah, please, what does that equal sign mean?" A speaker, familiar with Wigner would have smelled danger, but this speaker responded confidently, "It means the left side equals the right side." Wigner rubbed his forehead gently and continued, "Ah, yes ... but how?"

"I don't understand what you mean."

"Ah, are they equal as numbers, or are they equal as functions?" After a somewhat lengthy pause during which the speaker studied the equation in question he replied, "They are equal as functions." Wigner persisted, "Ah yes, but on the left you have a function of p and on the right a function of q. How is that possible?" After a second pause and more scrutinizing of the equation the speaker responded, "They are equal when p equals q."

With his index finger to his mouth, in a somewhat shy pose, Wigner continued, "Ah, well ah, why don't you write that?" The speaker did and to everyone's great embarrassment it was soon evident that the whole idea being presented was wrong.

That Bargmann was both thoroughly honest and also had a humorous side to him is demonstrated by the following. When a young Gerard G. Emch arrived at Princeton to start his postdoctoral fellowship he came with a freshly worked out theorem to present to Bargmann who immediately produced a counter example. Deeply disappointed, Emch went home to study the example. At 3 A.M. his telephone rang. Bargmann was on the other end and told him, "I thought you'd still be up. Go to bed and get some sleep. I found an error in my counter example. We can discuss it in the morning."

One of my classmates, Albert Clark, was Wigner's graduate student and related the following. Several Princeton professors, including Wigner, had cottages at Cape Cod. One autumn, as the maples were resplendent in their proud reds, Wigner invited Albert as well as his father to his cottage for a weekend. The ground was a riot of colors and Wigner raked the leaves into a pile and lit them.

The pungent odor was just spreading when an unexpected breeze sprang up and drove a portion of the suddenly blazing leaves against the cottage. Wigner grabbed a pail and rushed into the cottage. Albert's father spied a garden hose, turned on the faucet and doused the flames. With the danger over, Albert and his father went into the cottage to see what was keeping Wigner only to find the eminent professor at the kitchen sink busy scrubbing the pail.

After I started to teach physics, I found that my students also enjoyed hearing some of the anecdotes about the people whose ideas they had to study. As a consequence I livened up my lectures on quantum mechanics with stories about Bohr, Heisenberg, Schrödinger, Dirac, and Born. In a similar manner, Einstein's numerous foibles and memorable utterances helped to make relativity lectures less austere.

I believe that making such personalities less remote, showing their idiosyncrasies and making them more like the rest of us answers a need that arises not, as in the case of politicians, from the fact that they wield power, but rather from the fact that their abilities are often so much more developed than our own. Consider the effect on us of the story of Sir Isaac Newton (1642 – 1727) and the Brachistochrone problem.

After Jacques Bernoulli (1654 – 1705) had solved the famous Brachistochrone problem he challenged the natural philosophers and mathematicians of Europe to find the solution within a year. This problem required the mathematicians to find a curve that connects two points, separated both horizontally and vertically, such that if a bead slides under the action of gravity and without friction from the higher to the lower point, the time taken is a minimum. It is a problem in what is today known as the calculus of variation. The mathematical machinery to handle this had just been invented by Bernoulli when he solved the problem. He therefore thought that no one else could solve it. Now Newton, because then master of the mint, had not been active in science for some time and did not hear of the problem until the year was almost up. However, upon addressing the problem he solved it in an extremely elegant fashion and in a very short time, using purely geometric means. Today after three-hundred years of mathematical progress no simpler solution exists. The modern solutions, using the calculus of variation, require several pages to write out. Newton's solution required only one page and was purely geometrical. When Bernoulli was presented with the unsigned solution he had no doubt as to who had created that marvelous piece of work. He commented, "I recognize the lion by his paw." Incidentally, the solution is a portion of a curve called a "cycloid".

A typical joke that physicists like to tell about themselves goes as follows. A lawyer, an accountant and a physicist are having a discussion. The question arises as to what is better, to have a wife or a mistress. The lawyer argues that a wife is better since this lends stability. The accountant claims it is better to have a mistress since this poses fewer commitments. The physicist argues for having both. Since if he is not at home, his wife thinks he is with his mistress and if he is not with his mistress, she thinks he is with his wife. This leaves him free to spend more time in his lab.

Another apocryphal story that has frequently made the rounds but—ignoring the obvious hyperbole—is even more revealing of the approach a theorist may take is the following. An experimentalist approaches a theorist with a graph of experimental data and asks him if he can explain it. The theorist studies the graph for a minute or so and replies, "sure". He then proceeds to fill several blackboards with equations and turns to the experimentalist, "There!" The experimentalist looks at the graph in the theorist's hand and tells him, "You have the graph upside down." The theorist looks at the graph, turns it over and after a pause tells his colleague, "You should have told me sooner. This is even easier to explain."

The story of Heisenberg caught speeding is also clearly apocryphal. The policeman who stopped him approached his car and asked, "Do you know how fast you were going?" Heisenberg replied, "No, I don't, but I know exactly where I was."

Stories such as this bring out aspects of the character of a theorist, such as Newton, better than any abstruse description of his works. The abilities of a Newton are so far beyond those of the common man that they seem almost superhuman and on occasion might lead one to suppose, as was done tongue-in-cheek, on one occasion by Philip Morrison (1915 – 2005) that "these people are from Mars".

Most of the founders, of what is called "Modern Physics", are now dead. Their names will occur and re-occur throughout this book since the greatest minds contributed to more than one important idea in physics. My generation was the last to mingle with the last of them and to record some of these memories before this generation also dies out is timely. I have not any pretensions to having produced a "scholarly" work. Some of the stories recorded here are Gossip, or physics folklore. I have tried to verify as many of the stories as I could, but I have not hesitated to include all stories that I heard and noted down on scraps of paper. Sometimes I had to rely on memory since it would have been rude to take out a pencil and paper and immediately write down what the speaker said.

The founders of "Modern Physics" created ideas sometimes so counter-intuitive and outrageous that science fiction pales by comparison. Yet their ideas describe nature to an accuracy inconceivable a few decades ago. Some of the magic of these ideas is what I want to illustrate, as well as the deep mystery that physics poses. Why is it that by studying the behavior of the tiniest parts, the building blocks of the atoms, are we then able to put all the parts back together and make correct inferences about galaxies or even the whole universe? Indeed, as Einstein remarked around 1930, "The most incomprehensible fact of Nature is the fact that Nature is comprehensible."

Chapter 2

Thermodynamics: Founders and Flounderers

"Thermodynamics is a funny subject. The first time you go through it, you don't understand it at all. The second time you go through it, you think you understand it, except for one or two small points. The third time you go through it, you know you don't understand it, but by that time you are so used to it, it doesn't bother you anymore." Arnold Sommerfeld 1950

In 1875 when the seventeen-year old, Max Planck (1858 – 1947), applied to Professor Gustav Kirchhoff (1824 – 1887) of the University of Berlin to study physics with him, the eminent professor told him quite frankly, "Why do you want to come into physics? All is done and understood." Since Planck did not think he had enough talent for his other choice, a career in music, he went against the professor's advice and continued with physics.

At the University of Berlin, Planck fell in love with thermodynamics. It did not require any such abstract concepts as atoms and was built on solid experimental foundations derived from the practical problem of designing better steam engines. It was based on two seemingly innocuous postulates.
1. Energy is conserved. In other words, it is impossible to build an engine that does nothing but create energy. Stated another way, it is impossible to build what is called a perpetual motion machine of the first kind.
2. It is impossible to build an engine that does nothing but take heat from a source at one temperature and converts all of the heat into mechanical energy. This is called a perpetual motion machine of the second kind. As Lord Kelvin (1824 – 1907) phrased it in 1848, "It is impossible by means of inanimate material agency to derive mechanical effect from any portion of matter by cooling it below the temperature of the coldest of the surrounding objects."

A somewhat irreverent statement of the first two Laws of Thermodynamics is:

1) You can't win.

2) You can't even break even.

The laws of thermodynamics were derived from practical observations made while designing better engines. Put into mathematical form they have far reaching consequences that extend beyond the phenomena they were intended to summarize. Among other processes, thermodynamics allows us to compute the maximum efficiency of any engine, from an internal combustion engine to a nuclear power plant. Of course real engines are even less efficient than these ideal ones in the theory, but the theory gives realistic goals to aim for.

At the end of the 19th century only a few physicists actually used what was considered the "abstract concept of atoms" in their calculations. Planck also wanted none of this abstract stuff; he wanted to do solid science. In that regard he would, as a mature scientist, applaud the (1892) statement by Heinrich Hertz (1857 – 1894) that, "The rigor of science requires that we distinguish well the undraped figure of nature herself from the gay-colored vestments with which we clothe her at our pleasure." He would also, initially, disagree with the great Ludwig Boltzmann (1844 – 1906) who strongly advanced the notion of atoms and in 1895 had replied to Hertz, "But I think the predilection for nudity would be carried too far if we were to forego every hypothesis. Only we must not demand too much from hypotheses. ... Every hypothesis must derive indubitable results from mechanically well-defined assumptions by mathematically correct methods." Near the end of the nineteenth century, atoms were far from accepted and definitely not yet 'in' as a valid scientific concept. On the other hand, thermodynamics was considered to be a solid science. This is understandable since it sprang from considering some very practical problems.

One of the inadvertent founders of thermodynamics was a most colorful character, Benjamin Thompson (1753 – 1814), an American. At the tender age of 19 he married a wealthy widow, age 33 and so could move in higher social circles. However, he was a staunch Empire Loyalist and his open support of the British crown soon forced him to abandon his wife and infant daughter and flee from Boston. After spying for the British forces he moved to England while the American revolution was still in progress. After this war ended he went to Germany where he served the Bavarian king as minister of war and in other offices. As a reward, the king gave him the title "Count Rumford".

Count Rumford performed his duties to the Bavarian king so well that to this day, in the city of Munich, there are two great monuments to him: a massive statue and a very famous park containing numerous beer gardens. The Bavarian king established the "Englisher Garten" at Count Rumford's suggestion. Later Rumford married the widow of the great chemist Antoine Lavoisier (1743 – 1794) who had been beheaded during the French Revolution.

While in Munich (1779 – 1799), Count Rumford assisted in the boring of cannons. The procedure used was to rigidly mount a huge brass cylinder and drive a gigantic drill rotated by the force of four horses into the cylinder. Rumford noticed that during the process, workmen kept pouring buckets of water on the future cannon to keep everything from overheating. This aroused his curiosity. He used a dull drill and continued pouring water onto the brass cylinder only to discover that he could bring the water to a boil and continue to do so as long as the work of drilling continued. This convinced him that heat was not a caloric fluid, as was currently believed, since there was no limit to the amount of heat that could be extracted from a chunk of metal. Rumford drew the correct conclusion: the mechanical work of boring produced heat. He interpreted this to mean that work and heat are simply different forms of energy.

In 1842 a German physician, Julius Rupert Mayer (1814 – 1878), published an estimate of this "mechanical equivalent" of heat. On a trip to the tropics as a ship's surgeon he noticed, during routine blood letting, that the sailors' venal blood was much brighter red there than in the colder climate of his home town. He concluded that this was because due to the higher temperature, less oxidation had taken place in their bodies since some of the heat required by the body came from the tropical heat. So, like Rumford, he hypothesized that heat and energy were the same thing. Using available data on specific heats he computed this mechanical equivalent of heat and obtained a value in reasonable agreement with modern values.

A few years later controversy arose over priority when James Prescott Joule (1818 – 1889) published his measurements of the mechanical equivalent of heat. Mayer's results were also ridiculed, especially in a newspaper article that stated that energy conservation was a joke. This was too much for the emotionally stressed Mayer. One night he jumped from his second floor bedroom window in a suicide attempt, but survived and was committed for almost two years to a mental institution. The children of his hometown of Heilbronn referred to him as *"Der närrische Mayer"* (the crazy Mayer).

In 1858, honors from various universities started to arrive and eventually even elevation to the nobility with the addition of "von" to his name. Mayer's work finally received the recognition it deserved when the great John Tyndall (1820 – 1893) wrote, "We've all taken sides in fights that should never have occurred to begin with. In the firmament of science, Mayer and Joule constitute a double star, the light of each being in a certain sense complementary to that of the other." A grateful Mayer wrote to Tyndall, "Your kindness impresses me all the more [since I have], for many years, been forced to habituate myself to a precisely opposite . . . treatment."

James Prescott Joule's father was a rather well-to-do brewer. The young Joule's tutor was the famous scientist John Dalton (1766 – 1844) who emphasized the need for meticulous work in measurement. As a boy Joule had an interest in steam locomotives and also experimented with electricity. In a series of tests on a servant girl, he gave her ever increasing jolts of electricity until she passed out. At this point he ceased his electrical experiments.

James learned of Rumford's experiments with cannons and, a few years after Mayer's publication, set out and measured the mechanical equivalent of heat by various means. His experience in precise temperature measurements, as required by his father's brewer's craft, were a great asset. His measurements were very precise. The present unit of energy, the joule, is named after him. The controversy over priority remains to this day.

This was the beginning of the science of thermodynamics, a branch of physics that was firmly established by the researches of the nineteenth century.

Incidentally, beer has played an important part on more than one occasion in the history of physics. Donald A. Glaser (1926 –) invented the bubble chamber after watching bubbles rise in a glass of beer. This is an important tool in high energy physics experiments. Also, Carlsberg breweries funded the Carlsberg mansion in which Niels Bohr lived.

That energy is conserved had already been accepted by many savants. In 1775, the Paris Academy of Science announced, "The Academy has resolved, this year to examine no longer any solutions to the problems on the following subjects: The duplication of the cube, the trisection of the angle, the quadrature of the circle, or any machine claiming to be a perpetuum mobile." [1] This last item refers to conservation of energy.

The father of thermodynamics was a French engineer, Sadi Carnot (1796

---

[1] Histoire de l'Accadémie des Sciences, Année 1775

– 1832). Always of a frail constitution, cholera deprived the world of his talents at an early age. He was not only one of the first to enunciate the law of the conservation of energy (First Law of Thermodynamics), but also formulated correctly the Second Law of Thermodynamics. It is therefore, not surprising that in France, the law of conservation of energy (the First Law of Thermodynamics) is called *"le principe Carnot"*. A later expert in thermodynamics, Max Planck, had this to say about Carnot. "He has unquestionably the merit of having given the first evaluation of the mechanical equivalent of heat." The reason for this statement is that Carnot described very detailed experiments to measure the mechanical equivalent of heat.

The further development of thermodynamics included many famous individuals, but foremost among these was a Yale professor, J. Willard Gibbs (1839 – 1903), whose initial contract stipulated "without salary". It was only after Johns Hopkins University offered Gibbs a rather attractive position that Yale offered him a salary, about two-thirds of what Johns Hopkins had proposed. As one of his colleagues wrote to him to persuade him to stay at Yale, "Johns Hopkins can get on vastly better without you than we can. We can not." Gibbs remained at Yale.

Gibbs was very rigorous, even rigid, in his thinking and behavior. According to one of his students, E. B. Wilson, Gibbs always lectured above the heads of his students and consistently refused to teach undergraduates at all. He knew his students did not follow him but did not alter his style on that account, having a definite idea how the subject should be presented. He once told Wilson that in all his years of teaching he had had only six students sufficiently prepared in mathematics to follow him.

Unlike modern physicists who frequently publish too often and too soon, Gibbs labored for many years, until he had cast thermodynamics into a coherent and complete theory. He then published his entire work, as a book, which to this day is still one of the clearest expositions of this subject. Ironically, he was relatively unlauded in his own country, although much appreciated among scientific circles in Europe, especially Germany. Max Planck spoke of him thus, "... whose name not only in America but in the whole world will ever be reckoned among the most renowned theoretical physicists of all times." Also Walther Nernst (whom we meet later) paid for a marble memorial to Gibbs, even though they had never met. Gibbs' writings were sufficient to inspire such admiration.

Gibbs also invented a notation which is a boon to every modern engineer: vector notation. This way of writing vectors allows great insight into and simplification of such diverse fields as electromagnetic theory and fluid mechanics.

There is one story that may illustrate to what extent J. Willard Gibbs was underestimated at his own university and why it took such a long time for him to be promoted to full professor. During a visit to Cambridge University, the president of Yale inquired about possible people to promote at Yale. The famous Scottish physicist, James Clerk Maxwell (1831 – 1879) immediately suggested Gibbs. At this time there was also a socially rather prominent individual, named Alan Gibbs, at Yale. Thus, the president replied with pleasure. "Oh, you mean Alan Gibbs."

"No! No!" answered Maxwell; "Willard Gibbs."

"Well, but he is a nobody. He just sits in his room and writes," came the president's disconsolate reply.

During the nineteenth century, as today, many philosophies abounded, but British empiricism dominated much of science. One of the results of this extreme empiricism was that the concept of atoms, which was already starting to prove itself very useful, was still considered to be bad science. The argument went along the following lines. Atoms are never observed and as unobservable quantities should be excluded from physical theories since the purpose of science is to describe the real world around us and not to speculate about unobservable entities. This was a Catch 22 situation because, before convincing evidence for atoms could be found, calculations had to be performed using atoms in order to determine how to best detect them. But this went contrary to accepted doctrine. One who had the courage to do just that was a tragic genius, Ludwig Boltzmann (1844 – 1906). The obstacles he faced were enormous.

Boltzmann was the younger colleague of Josef Loschmidt (1821 – 1895), the first person to calculate the size of a molecule by using the kinetic theory of gases. In fact, what is called Avogadro's number in the English speaking world is called Loschmidt's number in the German speaking world. This is what James Clerk Maxwell said about this paper. "Loschmidt has deduced from the dynamical theory the following remarkable proportion: As the volume of a gas is to the combined volume of all the molecules contained in it, so is the mean path of a molecule to one-eighth of the diameter of a molecule."

Loschmidt also invented the markings for double and triple bonds of carbon and suggested the structural chemical formulas for a host of important molecules. His results were ignored for decades by the world's chemists.

Years later, Boltzmann visited Loschmidt after the latter's retirement and was shocked by the indigence of this great scientist. In Loschmidt's obituary he wrote, "This is how Vienna treats its great men."

Boltzmann went much further than his colleague and almost single-handedly created a new discipline, *Statistical Mechanics*, and in so doing unified the study of systems of atoms or molecules with thermodynamics. His main idea was that everything is made up of atoms and the very exact laws of thermodynamics result from the random collision of these atoms because in all macroscopic phenomena the number of atoms involved is enormous. This means that averages over these huge numbers lead to near certainty.

Ludwig Boltzmann as a young man.

To understand why this is so, consider flipping an unbiased coin. It is impossible to predict the result of a single toss. However, it is possible to make predictions about the average of a large number of flips. The prediction that, *on average*, one should get 50% heads gets better and better with the number of flips. In fact there exists a mathematical proof that the prediction of 50% heads improves with the number of tosses —the error in the prediction decreases like the reciprocal of the square root of the number of tosses. What this means is that for 100 tosses, the fractional error is about $1/10$, or 10%. For $10,000$ tosses about $1/100$, or 1%, and for $1,000,000,000,000$ tosses, the error is only about $1/1,000,000$ or 0.00001%. Even this error is large compared to the error involved with atoms since typically about $10^{24} = 1,000,000,000,000,000,000,000,000$ of them are involved so that the typical fractional error is of the order of $10^{-12} = 0.0000000001\%$.

Of course the empiricists did not take Boltzmann's assaults on their beliefs lightly. Many of his contemporaries criticized him severely, not only for what he said, but also most unfairly for "lacking elegance". To these remarks he replied, "Elegance is for shoemakers and tailors."

Some of the best minds of the nineteenth century grappled with Boltzmann's approach and found what appeared to be serious logical flaws. Again single-handedly (or better yet, single-mindedly) Boltzmann overcame all these difficulties and used the criticisms to improve and strengthen his theories so that his famous Boltzmann equation remains today, over a hundred years later, the best description available for non-equilibrium systems.

One of the people who found an apparent flaw in Boltzmann's work was Ernst Zermelo (1871 – 1956), a pure logician, after whom this result is called the "Zermelo Recurrence Paradox". He showed that all systems eventually return (recur) to almost exactly the same state. Thus, Boltzmann's proof that systems tend to a state of equilibrium had to be wrong. Boltzmann was able to show that this apparent paradox was due to a failure on Zermelo's part to distinguish between abstract logical and actual physical systems.

Whereas it is true that all mathematical systems must eventually recur, it is also true that, as Boltzmann showed, this recurrence time is much longer than even the age of the universe whereas the time to come to equilibrium is extremely short. Thus, Zermelo's proof, although logically quite correct, was totally irrelevant for physical systems.

As stated, Zermelo was a logician and while still only a Privatdozent (meaning that he had the right to teach without salary) at the University of Göttingen presented, according to Pauli, a rather irreverent version of the Russell's Paradox to one of his classes on mathematical logic. At the time the mathematics department at Göttingen was ruled by Felix Klein. Zermelo presented his class with the following problem. "All mathematicians in Göttingen belong to one of two classes. In the first class are those mathematicians who do what Felix Klein does, but what they dislike. In the second class are those mathematicians who do what they like, but what Felix Klein dislikes. To what class does Felix Klein belong?"

According to Pauli, none of the students was able to solve this problem. Zermelo then gloated, "Meine Herren, it's very simple. Felix Klein isn't a mathematician." Pauli finished the story with, "Zermelo was not offered a professorship at Göttingen."

By 1905, Zermelo had achieved some fame. He had worked on one of the famous 23 unsolved fundamental problems of mathematics, originally posed by David Hilbert, and given a proof of what mathematicians call "the well ordering theorem". This was enough to bring his name to the attention of the mathematics world. In 1910, he accepted the chair of mathematics at Zürich, but resigned this position in 1916 due to poor health and returned to the Black Forest in Germany where he remained for ten years. In 1926, he was appointed Honorary Professor at Freiburg im Breisgau. He resigned this position in 1935 as a protest against Hitler's regime. In 1945, he requested to be reinstated and again became an Honorary Professor at Freiburg im Breisgau in 1946.

Zermelo carried his logical analysis to extremes. He even analyzed everybody's statements. But he also had his slightly paranoid side. Once while he and several colleagues were at their *mathemathische Stammtisch* (literally "mathematical tribal table" that is, the table reserved for the regular mathematical guests) in the Schwarzen Bären in Göttingen, the meals were served and Zermelo looked around the table and pointing to one of the dishes protested. "There, of course, you have a much bigger portion and a much bigger slice than I have. They always treat me badly." Whereupon the plates were offered for exchange and the offer was graciously accepted.

Among these members of the Göttinger Stammtisch at the Schwarzen Bären was also Max Abraham (1875 – 1922), who had studied with Planck. He was very attached to the ether theory. So much so that in his obituary Max Born and Max von Laue wrote, "He loved his absolute aether, his field equations, his rigid electron just as a youth loves his first flame, whose memory no later experience can extinguish."

His contemporaries thought Abraham to be very sarcastic and thus he cultivated many enemies which may explain why he was not promoted. On one occasion, after the *Göttingen Academy of Sciences* had accepted a paper which Abraham considered wrong, the author of this paper, a young man named Madelung, joined the Stammtisch and was introduced to Abraham. This fine gentleman greeted him with the words, "Hallo, you have cheated our learned Academy nicely by persuading them to publish your paper. I congratulate you." Actually Madelung's paper on the absorption of infrared radiation by crystals was correct.

Another eminent critic of Boltzmann's work was Henri Poincaré (1854 – 1912). The important point is that Boltzmann had to overcome tremendous opposition to have his work accepted as the great work that it was.

The strain of all this work, however, together with the fact that he was turning blind and suffering from severe depression caused this genius to finally release himself from suffering by ending his life. While on vacation near Trieste he had an argument with his wife. She and their daughter left the house to go for a walk and when she returned, found her husband dead at the age of sixty-two.

The city of Vienna, home of many great men unrecognized until after their death, has in one of its cemeteries a tombstone bearing the inscription

$$S = k \log W \quad .$$

Beneath it lie the remains of Ludwig Boltzmann who opened the door to our understanding of macroscopic physics on the basis of microscopic or atomic dynamics. Thanks to him, statistical mechanics was firmly established as a rigorous and powerful discipline and the concept of atoms became respectable among scientists. Incidentally, the equation on his tombstone is due to Planck, not Boltzmann; it is simply a direct consequence of Boltzmann's work. The constant k was first introduced by Planck who called it Boltzmann's constant. That name has survived. When Planck was asked what he thought about the fact that this constant was not named after him he replied, "One constant is enough for me." The Boltzmann equation, which has withstood rigorous examination for more than a century, is not inscribed on Boltzmann's tombstone.

In spite of the fact that Boltzmann suffered from depression much of his life and found it extremely difficult to lecture he has, through his writings, revealed a rather impish soul. His account of his 1905 visit to California which translates as *A German Professor's Trip to El Dorado* is filled with humor and it is difficult to believe that only a year later he took his life.

In this account he voices his complaint about prohibition. Soon after his arrival in the New World he was treated to "an excellent dinner of fresh oysters". However, he was forced to drink something other than beer or wine, a procedure he considered barbarian. His main complaint about Berkeley also dealt with prohibition at that establishment, which forced him to drink water with the reaction that he had "to keep my clothes on all night to reach the toilet in time". Soon, however, he discovered the existence of an excellent wine merchant in nearby Oakland and thus was able to survive. His comment on America's "Noble Experiment" was, "The temperance movement is well on its way to giving the world a new species of hypocrisy."

Boltzmann however was not alone in advancing the notion of atoms. Hermann von Helmholtz in an address to the *Akademie der Wissenschaften zu Berlin*, in February, 1882 had this to say about the Second Law of Thermodynamics. "Unordered motion, in contrast, would be such that the motion of each individual particle need have no similarity to its neighbors. We have ample grounds to believe that heat-motion is of the latter kind, and one may in this sense characterize the magnitude of the entropy as the measure of the disorder."

In a similar manner James Clerk Maxwell explained to Lord John William Strutt Rayleigh, Third Baron of Terling Place (1842 – 1919) how he regarded the validity of the Second Law as being very probable. "Moral: The Second Law of Thermodynamics has the same degree of truth as the statement that if you throw a tumblerful of water into the sea, you cannot get the same tumblerful out again."

At the end of the nineteenth century, Classical Mechanics, Thermodynamics, and Electromagnetic Theory seemed able to explain the entire physical world. So, it was not without cause that physicists thought that "all is done and understood". However, soon this was revealed as hubris. Although barely noticed, cracks were already beginning to show in this magnificent structure called physics. These cracks grew progressively larger until the whole structure crumbled, only to be replaced with a new edifice in which the old structure is still recognizable, but greatly changed and much richer in design.

Chapter 3

Cracks Appear in Classical Physics

"There is nothing new to be discovered in physics now. All that remains is more and more precise measurement." Lord Kelvin to the British Association for the Advancement of Science in 1900.

As the end of the nineteenth century approached, physicists were justifiably proud of their science. It seemed that they had found the theories necessary to explain all physical phenomena. On Friday, April 30, 1897, a couple of years after Planck commenced his studies at the University of Berlin, the great British physicist, Sir Joseph John Thomson (1856 – 1940), announced to the regular evening meeting of the Royal Institute that he had discovered a particle smaller than an atom. Considering the fact that the very existence of atoms was still a controversial topic, this announcement was such a fantastic event that one of the prominent colleagues present thought it was a hoax. JJ, as Thomson was known, was not surprised at that reaction since he himself had found his discovery difficult to accept.

Actually, some of Thomson's results had been preempted by the Irish physicist, George Johnstone Stoney (1826 – 1911) who had introduced the name "electron" in 1891 in a paper in *The Scientific Transactions of the Royal Dublin Society.* The word electron comes from the Greek word for amber since rubbing amber produces an electrostatic charge. Stoney was motivated to introduce a set of "natural units independent of human existence" and cited a statement by Hermann von Helmholtz. "Now the most startling result of Faraday's Law is perhaps this. If we accept the hypothesis that the elementary substances are composed of atoms, we cannot avoid concluding that electricity also, positive as well as negative, is divided into definite elementary portions, which behave like atoms of electricity. As long as it moves about on the electrolytic liquid each ion remains united with its electric equivalent or equivalents. At the surface of the electrodes decomposition can take place if there is sufficient electromotive force, and then the ions give off their electric charges and become electrically neutral."

Thomson's results had also been preempted by Emil Wiechert (1861 – 1928). On January 7, 1896 Wiechert stated before the Königsberg Physics-Economics Society that the mass of the moving particles in the cathode ray beam seemed to be 2000 to 4000 times smaller than that of the hydrogen atom, the lightest of the known chemical atoms. In September 1897, he again reported to the German Society of Scientists and Physicists in Braunschweig, "The mass of the cathode ray particles is between 1/2000 and 1/1000 of a hydrogen ion."[1] On the other hand, he also stated, "So far as modern science is concerned, we have to abandon completely the idea that by going into the realm of the small we shall reach the ultimate foundations of the universe. I believe we can abandon this idea without any regret. The universe is infinite in all directions, not only above us in the large but also below us in the small." Physicists—in particular Wiechert—could not easily abandon the idea that matter is infinitely divisible.

The end of the nineteenth century was a time of great discoveries. In Germany, Professor Wilhelm Konrad Röntgen (1845 – 1923), winner of the first Nobel Prize in physics in 1901, had only a few years earlier discovered X-rays. Regarding this remarkable discovery he wrote in a letter dated Feb. 8, 1896 to his friend Zehnder, "To my wife I mentioned merely that I was doing something of which people when they found out about it would say, 'Röntgen has certainly gone mad'."

He had published his famous paper on X-rays in 1895 and received a congratulatory telegram from the Kaiser, "I praise God for granting our German fatherland this new triumph of science." On January 13, 1896 Röntgen presented himself at court and demonstrated, as requested, his X-rays to the Kaiser. This involved giving a demonstration, a lecture, a dinner with the Kaiser, and receiving the decoration, *Kronen-Orden 2. Klasse.* Röntgen was understandably nervous. He wrote to his friend, "I hope I shall have 'Kaiser luck' with this tube, for these tubes are very sensitive and are often destroyed ... and it takes about four days to evacuate a new one." The demonstration went off without a mishap. The press raved about Röntgen's discovery; newspapers not only published all sorts of X-

---

[1]These early measurements were actually very good. E. Wiechert (1896): 1.26×10^{11} C/kg. J.J. Thomson (1897): 0.7×10^{11} C/kg. Walter Kaufmann (1897 – 98): 1.77×10^{11} C/kg. Modern: 1.76×10^{11} C/kg.

ray images, but went on flights of fancy as to the possibilities that X-rays provided. This was carried to such extremes that some women even bought lead-lined underwear to protect their modesty from the view of prurient physicists.

Röntgen well illustrates how reluctant physicists were to accept new particles. He also did not accept the discovery of this new particle which JJ had labeled the electron and even forbade the use of the word electron in his laboratory. He was convinced that one had a moral duty to science not to talk in public about things that were not totally confirmed. His very strong moral convictions even carried over into his personal life. In 1918, near the end of World War One, Röntgen, by now in his seventies, was, along with the rest of the German people, slowly starving. Friends in Holland, aware of this, sent him packages of sugar and butter. However, he felt that it was not fair for him to consume these gifts and so he passed them on to other people.

Although the consensus among the physics community, as the nineteenth century drew to a close, was that, except for a few puzzling results, "all was understood" still more strange results were surfacing. In 1896 in France, Antoine Henri Becquerel (1852 – 1908) whose father and grandfather had also both been distinguished physicists, made a most startling discovery. He was investigating whether exposure to light caused certain phosphorescent minerals to increase their ability to darken photographic plates. Unfortunately, cloudy weather persisted for several days and he placed some unexposed photographic plates, carefully wrapped in dark paper, in a drawer and a lump of such a mineral on top of them. Much to his surprise, a few days later when he went to use these plates he found that they had been darkened precisely where the mineral had rested above them, even though the plates had been carefully wrapped to exclude all light. He was a sufficiently careful experimenter to realize that this might be significant and so discovered this highly penetrating radiation. This may be viewed as the beginning of nuclear physics, although it still took a while for atoms, let alone nuclei, to be accepted by the physics community.

Ironically, experiments performed even earlier, by the same Kirchhoff who had tried to discourage the young Planck from going into physics, soon showed that all was not well with physics and so led to a great upheaval that took more than a quarter century to resolve. In fact there were three results that were of a particularly disturbing nature: the photoelectric effect discovered by Heinrich Hertz in the middle of the nineteenth century, the Michelson-Morley experiment from the same period, and Kirchhoff's experiments dealing with blackbody radiation.

All of these results were totally inexplicable by the physics of the nine-teenth century and awaited ideas even more radical than the concept of atoms. Of course, these "anomalous" results were studied and, since they could not be explained, were relegated to the back burner to be revisited later.

When Kirchhoff told the young Planck that in physics "all is done", he failed to mention that he had just found that all bodies at a given temperature radiated roughly the same kind of light. This was not an entirely new discovery. It had been known in a rough sort of way by Josiah Wedgwood (1730 – 1795), the maker of fine ceramics, who had discovered that he could tell the right temperature of his kilns by looking at the color inside the kiln. Even much earlier, sword smiths had realized that they could judge, by its color, the right temperature of the metal for quenching. Kirchhoff, however, had gone much further and shown that this was a universal phenomenon and that the only thing that differed for specific materials was how much light was radiated, but not the quality of the light. So he defined a perfect radiator, called a "blackbody radiator" since materials covered with soot radiated better than shiny materials. Here was a problem to attract a physicist. Here was something universal and therefore obviously important. The problem: To find a formula to describe the radiation emanating from a blackbody at a given temperature.

Wilhelm Wien (1864 – 1928), using well-founded thermodynamic argu-ments, found a formula that agreed well with experiment for high frequen-cies, but failed for low frequencies.

On the other hand, Lord Rayleigh came up with a formula based on quite different but also firm theoretical foundations. For low frequencies or colors toward the red and infrared part of the spectrum, this formula agreed splendidly with experiment. However, for high frequencies, or colors toward the ultraviolet part of the spectrum the formula failed miserably. In fact it predicted the nonsensical result that the intensity of the radiation would increase rapidly with frequency so that if you looked at a stove you would be fried by ultraviolet, and even more so by higher frequency radi-ation. Physicists called this failure of the Rayleigh theory the "Ultraviolet Catastrophe".

Here were two formulas, but neither worked over the whole range of fre-quencies. Most physicists, that thought about this problem at all, believed that Wien's formula gave the correct description. They felt that Rayleigh's formula was somehow wrong, but could not find a flaw in his reasoning.

The next crack in the foundations of physics had been discovered even earlier by Heinrich Rudolf Hertz (1857 – 1894). In his short life, Hertz not only made important advances in both the theory and experimental verification of electrodynamics, but also discovered an effect that was to be one of the causes for the overthrow of classical physics. Hertz definitely had the skills for detailed experimental work. As a boy he had spent a lot of time at cabinet making and wood turning. His skill was such that the craftsman who had taught him was very disappointed when Hertz accepted a professorial appointment. "What a pity! That boy had the makings of a first-class woodturner."

In the 1880s, Hertz set out to see which of the two competing theories of electrodynamics: Maxwell's or Helmholtz's was correct. Although he leaned towards Helmholtz's theory, since Helmholtz had been his teacher, he showed unequivocally that Maxwell's theory was correct. To do this he produced the first radio (electromagnetic) waves and showed that they travelled at the speed of light and were refracted just like light as Maxwell's equations predicted.

When someone had asked Maxwell what electricity was good for he replied, "I do not know, but I'm pretty sure that Her majesty's government will soon tax it."

Regarding Maxwell's equations, Hertz later stated, "One cannot escape the feeling that these mathematical formulae have an independent existence and an intelligence of their own, that they are wiser than we are, wiser even than their discoverers, that we get more out of them than was originally put into them."

Even the great Hendrik Antoon Lorentz (1853 – 1928) was converted from Helmholtz's electromagnetic theory by Hertz's experiments, which he felt were "the greatest triumph that Maxwell's theory has achieved."

Incidentally, it was a description of this work in Hertz's obituary that led a rather spoiled young man, encouraged by his mother, to start experimenting with these waves. So it came about that Guglielmo Marconi (1874 – 1937) succeeded in producing the first wireless transatlantic signal in 1900.

What was it that Hertz had found that was such a puzzle? Being an extremely careful observer, Hertz had noted that when the flash of light from an electric spark hit a clean metallic plate it made the air near the plate more conducting. Although this was not relevant for his experiments with radio waves, he carefully noted the various aspects of this phenomenon, later called the "photoelectric effect". The problem with this effect was that it defied all attempts at an explanation. The explanation came almost fifty years later from a Swiss patent clerk who in 1905 published three papers that revolutionized physics. But before we get to that we have one more puzzle to address.

In the middle of the nineteenth century, Albert Abraham Michelson (1852 – 1931) set out to measure the speed with which the earth moved through the luminiferous ether. This experiment yielded results that were not explained until, as discussed in the next chapter, Special Relativity was invented. The ether was supposed to be an elastic solid existing everywhere, throughout all space, yet allowing other solid bodies to move through it without resistance. Light waves were thought to propagate through space as vibrations of this ether, sort of in analogy to the way vibrations of gases carry sound waves. The way to do Michelson's experiment was to measure the speed of light in the direction of travel around the sun, as our planet moves one way through the ether in the galaxy as well as in the opposite direction. To Michelson's surprise and the shock of the physics world he found the speed of light from the sun was the same, regardless of the direction in which the earth is moving.

Chapter 4

It's About Time and Space

"Everything that is relative presupposes something that is absolute, and is meaningful only when juxtaposed to something absolute." *Max Planck*

Scientific revolutions are much like political revolutions. During their initial phases they are ignored by almost everyone except the revolutionary activists and their conservative opposition. It is only when the revolution is full-blown or almost over that the rest of the world takes notice. Thus, it was at the beginning of the twentieth century in physics. There were flaws in the existing science, but most physicists could ignore them since there were really only the three flaws mentioned earlier, and these were rather esoteric little flaws. Similarly, there were flaws (social injustices) in many countries, but people could also ignore these. In both cases the revolutions that ensued left the world forever changed.

One of the revolutionaries in physics was William Thomson, later Lord Kelvin of Largs (1824 – 1907). Both William Thomson (Lord Kelvin) and his brother had been tutored by their father who was a professor of mathematics, first in Belfast and then in Glasgow. Lord Kelvin was a child prodigy and had enrolled in the University of Glasgow at age ten to study Natural Philosophy. He retired from the University of Glasgow at age 75 even though the trustees indicated that they would be happy to retain his services. He, however, put it, "No sentimentality, if you please. I have outlived my usefulness." The following year he enrolled in the university as a research student because he wanted to "keep his hand in". Thus, he became both the oldest and the youngest student to enroll in Glasgow University.

At the start of the twentieth century Lord Kelvin kept pointing to the flaws (he called them clouds on the horizon) of the physics of the time. The first manifesto by Kelvin was a paper published in 1901 entitled "Nineteenth century clouds over the dynamical theory of heat and light". Actually one of the clouds had already burst the year before as a thunder-storm creating a flash flood that eventually swept away much of the surface structure of nineteenth-century physics. But the cloudburst had gone almost totally unnoticed.

John William Strutt, Lord Rayleigh and Thomson William Kelvin, Lord Kelvin of Largs in Rayleigh's laboratory at Terling Place.

A few years later (1904) Lord Kelvin repeated his theme in his Baltimore Lectures stating: "The beauty and clearness of the dynamical theory, which asserts heat and light to be modes of motion, is at present obscured by two clouds .... The first ... involved the question, How could the earth move through an elastic solid, such as essentially is the luminiferous ether? The second is the Maxwell-Boltzmann doctrine of partition of energy." He had missed one cloud, but more of this later.

William Thomson was known for his self-confidence. As an undergraduate at Cambridge he was certain that he was "Senior Wrangler" (the student who scored highest on the Cambridge mathematical Tripos exam). After the exam, when the marks were posted, he asked his servant, "Oh, just run down to the Senate House, will you, and see who is Second Wrangler."

The servant returned and informed him, "You, sir!"

Lord Kelvin was also very steeped in the British tradition of concrete models. In his 1884 Baltimore lectures he confessed. "I never satisfy myself until I can make a mechanical model of a thing. If I can make a mechanical model, I can understand it. As long as I cannot make a mechanical model all the way through I cannot understand; and that is why I cannot get the electromagnetic theory."

Later in life after he had failed to find a mechanical model for electro-magnetism, he wrote in a letter to George Francis Fitzgerald (1851 – 1901) in 1896, "I may add that I have been considering the subject for forty-two years. I have been trying many days and many nights to find an explanation, but have not found it. ... I have not had a moment's peace since Nov. 28, 1846. All this time I have been liable to fits of ether dipsomania, kept away only at intervals by rigorous abstention from thought of the subject."

He was also a no-nonsense scientist who appears to have had little patience with the so-called soft sciences. "When you can measure what you are speaking about and summarize it in numbers, you know something about it. And when you cannot express it in numbers, your knowledge is of a meager and unsatisfactory kind. It may be the beginning of knowledge, but you have scarcely in your thought advanced to the stage of science." His conservative nature made him prone to unwarranted pronouncements as his statement in 1895, when he was president of the Royal Society, illustrates. "Heavier-than-air flying machines are impossible." He followed this in 1896 with, "I have not the smallest molecule of faith in aerial navigation other than ballooning. ... I would not care to be a member of the Aeronautical Society."

Uniformitarianism, was another of the ideas that Lord Kelvin would not accept. This is the idea that geological processes have always proceeded at roughly the same slow rate as we observe today. His reason was that this did not agree with his calculated rate of the earth's cooling which gave an age of the earth of only about 100 million years, much shorter than what geologists claimed. Charles Darwin's estimate that evolution had required a minimum of several hundred million years was also in serious conflict with Lord Kelvin's computation. To avoid conflict with Kelvin, Darwin omitted this from later editions of his *The Origin of Species*.

In his opposition to Uniformtarianism, Lord Kelvin was strongly and stridently, supported by his friend, Peter Guthrie Tait (1831 – 1901). This same Tait, in the third edition of his book *Quaternions*, had also attacked Gibbs' most useful development of the vector calculus. "Even Prof. Willard Gibbs must be ranked as one of the retarders of quaternion progress, in virtue of his pamphlet on *Vector Analysis*, a sort of hermaphrodite monster, compounded of the notations of Hamilton and Grassmann." To this day, physicists and engineers are grateful to Gibbs for the great advantages that vector calculus provides over other approaches.

The quaternions touted by Guthrie had been invented by Sir William Rowan Hamilton (1805 – 1865) on the way to a meeting of the Irish Royal Society. He was so pleased with his invention that he scratched it into the stone of Brougham bridge in Dublin where it still stands.

$$i^2 = j^2 = k^2 = ijk = -1.$$

Due to the very precise temperature measurements he carried out, Lord Kelvin's name is immortalized in the name for the absolute temperature scale whose unit is the "kelvin". He also formulated the Second Law of Thermodynamics. For him the key issue, in the interpretation of the Second Law of Thermodynamics, was the explanation of irreversible processes. He noted that if entropy always increased, the universe would eventually reach a state of uniform temperature and maximum entropy from which it would not be possible to extract any work. He called this the "Heat Death of the Universe".

It is less well known that he contributed extensively to the ideas that led to the modern refrigerator. In fact, one of the early refrigerator companies named its device the "kelvinator" to honor his achievements. What is very surprising, for someone as expert in thermodynamics as Kelvin, is that at one point he tried to demonstrate that the equipartition theorem[1] was invalid.

Incidentally, his wife, Lady Kelvin was a most "solicitous" spouse who kept a close eye on her husband. Walther Nernst has written that he and Kelvin met Madam Marie (Sklodowska) Curie (1867 – 1934) at a dinner party. Naturally they asked her about the element radium that she had just newly discovered. She told them that she had brought along a sample but that the lights were too bright to see it glow. So the three of them squeezed into the space between a set of double doors where it would be dark enough. But, before their eyes could get dark-adapted to see the glow, Lady Kelvin knocked on the doors. Perhaps Lady Kelvin's solicitude had

---

[1]See chapter 6

something to do with the fact that as a young woman, Madam Curie was extremely attractive.

Marie Sklodowska Curie as a young woman.

The main point in referring to Lord Kelvin's speeches is that, in spite of "the clouds obscuring the clarity and beauty of the dynamical theories", for most physicists it was physics as usual. Only for a few revolutionaries were serious problems looming. Even Lord Kelvin (although very aware of these clouds) stated in an address to the British Association for the Advancement of Science in 1900, "There is nothing new to be discovered in physics now. All that remains is more and more precise measurement." Later in life, the American physicist Albert Michelson echoed Kelvin, "The grand underlying principles have been firmly established. ... further truths of physics are to be looked for in the sixth place of decimals."

As everyone knows, when a ball is thrown at you the speed with which it arrives depends on whether you are standing still or running towards or away from the thrower. If you run away fast enough the ball will never catch up to you; you outrun the ball. It is also possible to outrun sound waves as planes do when they fly at supersonic speeds. To the nineteenth century physicist it was equally obvious that if one went fast enough one could outrun any wave, including light. To achieve such high speeds was, of course, technologically impossible, but the difference in the observed speed of light when moving in the direction of the light wave or against it was certainly something that should be measurable. This was precisely the purpose of the Michelson-Morley experiment. The result of this experiment produced the first dark cloud that Lord Kelvin referred to. To dissipate this cloud required no less than a new theory: *The Special Theory of Relativity.*

The story of special relativity might begin with the labors of one Albert Abraham Michelson, the son of Jewish immigrants to the Americas. He grew up in the mining camps of California and Nevada during the American civil war and showed such early promise that he received an overquota appointment to the US Naval Academy from President U. S. Grant. Eventually Michelson became an instructor at the academy and developed into the world's foremost master of measurements involving light. So much so, that in 1907 he was the first American to receive the Nobel Prize "for his precision optical instruments and the spectroscopic and metrological investigations conducted therewith".

His road to finding the flaw that led to special relativity began when he attended the *Montreal Meeting of the British Association for the Advancement of Science* where he met Lord Rayleigh. Later he also heard the Baltimore lectures delivered by Lord Kelvin, at that time still William Thomson. These two eminent physicists became his main scientific advisers and moral supporters, providing him with the much-needed encouragement for his work.

To understand the Michelson-Morley experiment, we have to return to Maxwell's theory of electromagnetism. As we already saw, Heinrich Hertz had tested that theory's prediction, regarding waves that travel at the speed of light, and found it eminently correct. The existence of these waves presented somewhat of a problem for the classical physicist with his completely mechanistic (British Empiricist) viewpoint. It was inconceivable to such a physicist that a vibration could exist without some material substance to do the actual vibrating. This function of vibrating or undulating was admirably served by a fictitious substance: the ether. The physicist's ether not only filled all of space and thus carried all electromagnetic waves, somewhat like air carries sound waves, but it also served to define a state of absolute rest. An object was, by definition, at rest if it was not moving relative to the ether.

Here is what James Clerk Maxwell wrote about the ether in his famous *Treatise on Electricity and Magnetism.* "In fact, whenever energy is transmitted from one body to another in time, there must be a medium or substance in which the energy exists after it leaves one body and before it reaches the other, for energy, as Torricelli remarked, is a quintessence of so subtile a nature that it cannot be contained in any vessel except the inmost substance of material things! Hence all these theories lead to the conception of a medium in which the propagation takes place, and if we admit this medium as a hypothesis, I think it ought to occupy a prominent place in our investigations and that we ought to endeavor to construct a mental representation of all the details of its action, and this has been my constant aim in this treatise."

The ether also played another role. The idea of a state of absolute rest was very important for Newton's dynamics and had been one of the big sources of debate between Newton and his contemporary, Gottfried Wilhelm Leibnitz (1646 – 1716). Leibnitz argued that space was not absolute, but that the position of an object could only be given relative to some other object. Newton, on the other hand, argued that it was possible to give the position of an object without reference to another object.

The success of Newton's laws had resolved the debate, by default, in Newton's favor. The ether served admirably to give Newton's absolute reference frame for the position of any object and thus satisfy the "pedants" who were still trying to find fault with Newton's concepts. For this reason, the physicists of that time were very happy to have the ether serve a dual purpose: as an absolute reference frame in addition as well as its primary purpose of carrying electromagnetic waves.

This primary function of the ether, to carry electromagnetic vibrations, caused difficulties almost from the start. These difficulties were mostly esthetic; but to physicists this was not a trivial matter. Most physicists agreed that such a successful theory should also be "beautiful". An example of these unpleasant difficulties was the fact that although the planets and other material bodies could move through the ether unhindered, the ether had to be completely incompressible. The ether could be twisted (sheared) without any resistance but could not be squeezed; it was harder than diamonds when squeezed. This strange incompressibility of the ether was necessary to explain the absence of longitudinal (like sound) electromagnetic waves. Although there was no logical reason why the ether could not behave like this, it did make the ether rather different from any other known substance.

Since the ether defined a state of absolute rest, Michelson proposed to measure the earth's absolute speed or, what is the same thing, its speed through the ether. The idea was simple, although technically extremely difficult.

As the sun moves through the ether and the earth revolves about the sun the earth must, part of the time, move in the same direction as the sun and, part of the time, in the opposite direction. Thus, all one has to do to measure the earth's speed through the ether is to measure the speed of light in a direction parallel to the earth's motion about the sun, and look for systematic variations over a period of a year. Or so it seemed.

Michelson and Morley planned such an experiment to measure the earth's absolute motion in great detail. They then carried out the measurement, with superb precision, only to have them seem to fail. No measurable difference in the speed of light over the period of a year, as the earth must have moved in the same direction as well as opposite direction to the sun, was observed. Although the experiment had an accuracy 50 times greater than what should have been required to measure the earth's speed, no difference was found. The situation was similar to a car going a hundred miles an hour and smashing into a train going fifty miles an hour only to find that the speed with which the train hits the car does not depend on whether the train is moving towards or away from it. The results contradicted all common sense. This was the famous Michelson-Morley experiment.

Michelson-Morley first performed their experiment in 1865. Fourteen years later Albert Einstein (1879 – 1955) was born in Germany. He grew up as a rather indifferent student. Eventually he even received a Ph.D. but no one expected too much of him. H. F. Weber was one of Einstein's teachers who recognized his abilities even though this student exasperated him. He once told him, "You are a clever fellow, but you have one fault. You won't let anyone tell you one thing." Later when Einstein failed his entrance exam to the Eidgenossische Technische Hochschule (ETH), Weber encouraged him not to give up. However after Einstein finished, Weber would not accept him as his assistant.

Similarly, Minkowski, one of Albert Einstein's teachers at the Polytechnical school in Zürich when he first saw Einstein's work on relativity uttered, "Imagine that! I never would have expected such a smart thing from that fellow." So, it was only through the intervention of a friend, that Einstein finally received a position as a clerk in the Swiss patent office. This may also be the reason why he suggested that, "Science is a wonderful thing if one does not have to earn one's living at it."

On the other hand as he later stated, landing a job at the patent offices was very fortunate for he had to examine many mechanical models to see if they would actually work. At the same time he met two people who were to remain his lifelong friends. They were Conrad Habicht and Marcel Solovine. Together they formed a small discussion group that they dubbed "The Olympia Society". The days at the patent office and in the company of his friends of the Olympia Society were Albert Einstein's happiest and most fertile days.

It was during his period as a patent clerk that Einstein solved the paradoxical results of the Michelson-Morley experiment: the problem of how the speed of light was independent of the speed of the source emitting the light as well as of the speed of the receiver.

How did Einstein resolve this difficulty? Very simply, he just stated that there is no difficulty; one should not expect a difference in the two speeds of light since no ether exists. In fact, according to Einstein, the only function of the ether was grammatical: to be the subject of the verb "to vibrate". In asserting this he seems to have been influenced by Ernst Mach. As he stated in Mach's obituary, "Even those who think of themselves as Mach's opponents, hardly know how much of Mach's views they have, as it were, imbibed with their mother's milk."

As a young man Einstein had wondered what a light beam would look like if one could move at the speed of light beside it. He concluded that it would still have to be a beam of light. This led him to postulate that the speed of light was independent of the speed of the source or the observer. He then added a second postulate: the laws of physics must be the same for all observers, regardless of the speed with which they are moving. These two seemingly innocuous postulates have immediate far-reaching consequences.

To see one of the consequences, consider a physicist with the following very accurate clock. He has a beam of light reflect back and forth between two parallel mirrors exactly 1.498 962 meters apart. This distance is chosen purely for convenience. In going back and forth the light beam travels twice this distance or 2.997 924 meters. Since the speed of light is $2.997\ 924 \times 10^8$ m/s, the time for a complete back and forth traverse is 10^{-8} or one hundred millionth of a second. Now consider a second physicist with an identical clock flying past the first one with their mirror clocks parallel to each other. As he flies past his colleague he sees that due to their relative motion, the light beam on his colleague's clock does not fly straight back and forth between the two mirrors, but carries out a zig-zag or saw-tooth path. In other words, his colleague's light beam has to travel further to go back and forth. Since the speed of light is constant this means that the light beam on his colleague's clock takes longer to travel back and forth. His friend's clock runs slow! Of course his friend looks across and also sees that the light beam on the second physicist's clock carries out a saw-tooth pattern and so is also slow. Both are correct in their observation. This first consequence of Einstein's two simple postulates is that time is not absolute, but relative.

In formulating his postulates, Einstein seems to have followed his own motto, "Everything should be made as simple as possible, but no simpler." Why was Einstein able to do this, and what motivated him? The answer to the first question is found in Leibnitz's debate with Newton. Almost two-hundred years before Einstein's birth, Leibnitz had asserted that only relative and not absolute motion had meaning. By banishing the ether, Einstein had vindicated and in fact reintroduced Leibnitz's viewpoint. What is surprising is that Einstein's motivation was of a quite different nature. As stated, he was not motivated by the Michelson-Morley experiment. In fact, Michelson's work is not referenced in any of Einstein's papers. This does not mean that Einstein did not appreciate the fact that the null result of the Michelson-Morley experiment was the prime evidence in support of Special Relativity.

Einstein's motivation, however, was of a quite different nature. It was purely esthetic. He was disturbed by an asymmetry of explanations that existed for the following situations.

If we have a magnetic field and a very long wire and we move the wire across this magnetic field, an electric current flows in the wire. This is called an induced current and is the basis of electric power generation. Similarly if the magnetic field is moved and the wire held still, a current also flows. Both cases are illustrations of exactly the same phenomenon. In fact, if the speed at which the magnetic field is moved is in the opposite direction but of the same magnitude as the speed with which the wire was moved, then the induced current is exactly the same in the two cases. After all, the two cases are indistinguishable unless there is some third object relative to which one can decide whether the wire or the magnetic field is moving. In Einstein's own words from 1952, "What led me more or less directly to the special theory of relativity was the conviction that the electromotive force acting on a body in motion in a magnetic field was nothing else but an electric field."

However, in the old ether theory, two different explanations had to be given for these two cases because the motions could be viewed against the background of a stationary ether and one could tell whether it was the wire or the magnetic field that was really moving. This, to Einstein, seemed to be a flaw in the theory even though the explanations gave correct answers. In removing this flaw, he did not change Maxwell's equations, which describe this phenomenon, instead he altered Newton's kinematics (which had stood for 200 years) and replaced it with Special Relativity.

Einstein and Leopold Infeld (1898 – 1968) in *The Evolution of Physics* (1938) had this to say. "The relativity theory arose from necessity, from serious and deep contradictions in the old theory from which there seemed no escape. The strength of the new theory lies in the consistency and simplicity with which it solves all these difficulties, using only a few very convincing assumptions. ... The old mechanics is valid for small velocities and forms the limiting case of the new one.

"In classical physics it was always assumed that clocks in motion and at rest have the same rhythm, that rods in motion and at rest have the same length. If the velocity of light is the same in all coordinate systems, if the relativity theory is valid, then we must sacrifice this assumption. It is difficult to get rid of deep-rooted prejudices, but there is no other way."

Einstein was not satisfied with a theory that merely reproduced the observed facts: "Agreement with experimental fact must not be sought in the initial steps of theoretical analysis but in the final results." Also Einstein did not explain the Michelson-Morley experiment; he simply made it the natural or expected result. This technique of turning the unexpected into the expected is a common approach in science and we shall encounter it again, especially with Niels Bohr and the *Old Quantum Theory*. It is a very powerful technique for creating new theories.

Although we have told this part of the story as if Einstein alone was responsible for the special theory of relativity, this is not correct. The Michelson-Morley experiment had, in fact, led Hendrik Antoon Lorentz, the grand old man of Dutch physics, to the transformation equations of special relativity. That is why these transformation equations are known as "Lorentz Transformations".

Furthermore, Henri Poincaré (1854 – 1912) had somewhat earlier than Einstein, and also less explicitly, stated the laws which govern the essential content of the special theory of relativity. He was probably governed by his own philosophy, "Science is built up with facts as a house is with stones. But a collection of facts is no more a science than a heap of stones is a house." He is also quoted as saying, "It is the simple hypotheses of which one must be most wary; because these are the ones that have the most chances of passing unnoticed." Thus, it is not surprising that with regard to the various contrived explanations of the Michelson Morley experiment he concluded, "If nature conspires over and over again to achieve a certain result, then it is no longer a conspiracy, but a law of nature. The speed of light is constant." It is therefore quite appropriate that he also has a part of Special Relativity named after him, the Poincaré Group.

The most transparent mathematical formulation of Einstein's theory was due to the same Hermann Minkowski (1864 – 1909) that was surprised by the "clever thing" that Einstein had done and after whom the four-dimensional space-time of Einstein's Special Theory of Relativity is named Minkowski space.

Minkowski was an excellent mathematician and teacher. One of his students, whom we meet later, was Max Born who took a course in Analysis Situs (now called Topology) from him. Minkowski began the course by stating, "A good introduction to this subject is the four-color problem. So far only third rate mathematicians have attacked this problem and there is no proof. While preparing these lectures I found a simple proof which I shall explain to you." So Minkowski started with the simplest concepts of topology and continued for several weeks. The students eagerly attended his class waiting to hear the history making "proof". One morning, after more than four weeks, when the students were assembling, a big thunderstorm broke and there was a great crash of thunder. Minkowski said with a serious mien, "Heaven's wrath is upon me for my conceit. My proof is also wrong." Then he started to chuckle. The whole thing had been a trick to keep the students' attention.

Once Minkowski had reformulated Special Relativity, he reported in 1908 to the 80th *Versammlung Deutscher Naturforscher und Ärzte in Köln* (Meeting of German Scientists and Doctors in Cologne). "Gentlemen, the views of space and time which I wish to lay before you have sprung from the soil of experimental physics, and therein lies their strength. They are radical. Henceforth space by itself, and time by itself, are doomed to fade away into mere shadows, and only a kind of union of the two will preserve an independent reality."

Nonetheless, Einstein developed independently and stated in the clearest manner the physical concepts of what is now the special theory of relativity and thus the credit is justifiably his.

There were, of course, conservative elements among the physics community that refused to accept Einstein's ideas. And later, during the dark days of the Nazi period, relativity theory was condemned as "Jewish physics" and forbidden to be taught in German schools. Nevertheless, physicists like Arnold Sommerfeld (1868 – 1951) and Max Planck (1858 – 1947) embraced

Einstein's ideas from the start and continued to espouse them even during the Nazi period.

In 1910 Planck wrote, with respect to Einstein and Special Relativity Theory, "If [it] ... should prove to be correct, as I expect it will, he will be considered the Copernicus of the twentieth century." Later, Planck revised Newton's laws to conform with relativity. In this he was guided by the action principle. For example, regarding conservation of energy and momentum he stated, "Over both of them reigns the principle of least action, which seems to control all the reversible processes of physics." He also gave a general proof of $E = mc^2$, as well as that entropy is a relativistic invariant. He furthermore derived the transformation laws for other thermodynamic quantities such as pressure, temperature, etc.

That Einstein's admiration for Planck was equally great was explicitly avowed: "Everything that emanated from his supremely great mind was as clear and beautiful as a great work of art; and one had the impression that it all came out so easily and effortlessly. ... For me personally, he meant more than all the others I have met on my life's journey."

Since Einstein dominated so much of the first half of the twentieth century in physics, there are many stories about him. The following story according to Walther H. Nernst (1864 – 1941), discoverer of the Third Law of Thermodynamics and Einstein's colleague at the University of Berlin, illustrates what a great wife the second Mrs. Einstein was. Einstein, a heavy smoker, was told by his doctor to quit cigars for a while, but as soon as his wife had turned her back he sent out the maid to get him a box. When Mrs. Einstein returned she noticed the heavy smoke in the flat but made no comment. She pretended not to notice, but looked at her husband with a very worried face saying that he looked very pale and asked whether he felt alright. Einstein got so worried that he quit smoking for some time. Nernst commented, "You see, she is an excellent wife, she saved his life."

On another occasion, after a colloquium, Nernst and Einstein argued fiercely without reaching an agreement. Shortly afterwards at another colloquium the same issue arose and Einstein now adopted a position very close to Nernst's original argument. This was too much for Nernst, to find Einstein now voicing his original opinion without the slightest acknowledgment of his conversion to it. Einstein's reply was typical of Einstein and the spirit of the colloquia at the University of Berlin, "But, really my dear colleague, Nernst, is it my fault that God created the world differently from what I thought three weeks ago?"

Like many famous physicists, Einstein was very interested in music. He loved playing the violin and was, in fact, quite good at it. However after he was already famous and gave a performance for charity, the local critic wrote, "Professor Einstein did not live up to his reputation."

That Einstein truly appreciated expert violin playing is evident from the following story. On an April evening in 1930 Einstein attended a concert at the Berlin Philharmonic conducted by Berno Walter. The performance featured the music of Bach, Beethoven, and Brahms. The guest violinist Yehudi Menuhin, as well as the orchestra, so overwhelmed Einstein that after the performance he rushed up to Menuhin, embraced him and exclaimed, "Now I know there is a God in heaven."

Einstein's fame brought him together with many celebrated artists. At one point he apparently played his violin for the famous cellist Gregor Piatigorsky, who was well known for his sense of humor. Afterwards either Einstein or someone else asked Piatigorsky what he thought of Einstein's playing. After a little stammering he said, "He played relatively well." There is another story along these lines. When again someone asked Piatigorsky about Einstein's playing he is supposed to have said, "Mister Einstein can't count."

As a young man, Einstein heard someone playing a Mozart sonata on the piano in the house in which he was staying. He asked his landlady where the music was coming from. She told him that it was the lady renting the upstairs apartment. Einstein grabbed his violin and charged up the stairs. Soon the people downstairs heard his violin accompanying the piano. When Einstein came back down he stated, "That is a wonderful old lady. I'm going to visit her often." Later the roomer from above appeared and told the landlady that she had been somewhat shocked by the young man bursting into her apartment and telling her to "keep playing".

Also, once while performing for some ladies who started to knit and whose needles clattered, Einstein packed up his violin with the words, "We should not disturb you at your work."

Einstein also used his violin to make a pun in German. At the home of Professor Otto Stern, Einstein got into quite a hefty dispute with another physicist. He ended the discussion by pointing to his violin and stating, "Why don't we retire to the music room? There we can play what you so very much enjoy, namely Händel!" The name Händel is a play on the German word "*handeln*", to barter or bargain.

When Einstein was already well settled in Princeton, a new postdoctoral fellow at the Institute for Advanced Studies was introduced to him. After an embarrassing silence, the young man told Einstein that he always carried a little black book with him in which he wrote down any good ideas he might have. He then asked Einstein if he kept such a book. The latter replied almost sadly that he did not since he had good ideas so seldom.

Someone who, in Prague, had made an appointment with Einstein to meet him on a bridge had forgotten and arrived two hours late only to find Einstein sitting on the side of the bridge. To his effusive apologies Einstein replied quite simply, "That's alright. Don't worry, I can think here just as well as anywhere else."

The following story was told by Gabriele Rabel. As a student she had met Einstein and, full of admiration of his quick mind, asked him how it was that he could immediately put his finger on the flaws of the theories of all those bright young men. Einstein smiled, "I cheat. You see, I know all those theories that those bright young men propose. I have thought of them all myself. So I know exactly where their flaws are."

There are also many notable quotes from Einstein. Below are a few.

"Common sense is the collection of prejudices acquired by the age of eighteen."

"Pure logical thinking cannot yield us any knowledge of the empirical world; all knowledge of reality starts from experience and ends in it."

"Most of the fundamental ideas of science are fundamentally simple, and may, as a rule, be expressed in a language comprehensible to everyone." The sentiments expressed in this statement are echoed by many physicists.

In a letter to Maurice Solovine in 1949 Einstein wrote, "There is not a single concept of which I am convinced that it will stand firm, and I feel uncertain whether I am, in general, on the right track."

Also, in a letter to the Queen Mother of Belgium he wrote "The strange thing about growing old is that the intimate identification with the here and now is slowly lost. One feels transposed into infinity, more or less alone, no longer in hope or fear, only observing."

"To punish me for my contempt for authority, Fate made me an authority myself."

"Not everything that can be counted counts, and not everything that counts can be counted." Contrast this with the statement by Lord Kelvin quoted earlier. "When you can measure what you are speaking about and summarize it in numbers, you know something about it. And when you cannot express it in numbers, your knowledge is of a meager and unsatisfactory kind. It may be the beginning of knowledge, but you have scarcely in your thought advanced to the stage of science."

Einstein was also aware of the ecological world around him. He once stated, "If the bee disappeared off the surface of the globe then man would only have four years of life left. No more bees, no more pollination, no more plants, no more animals, no more man."

That new ideas are difficult to accept is illustrated by some sentences in Michelson's book *Studies in Optics* published in 1927. In this book he shows how the Michelson-Morley and several other optical effects follow from Special Relativity. However he adds, "The existence of an ether appears to be inconsistent with the theory. ... But without a medium how can the propagation of light be explained? ... How explain the constancy of propagation, the fundamental assumption (at least of the restricted theory) if there be no medium?"

In 1931, four months before Michelson's death, Einstein finally met him in Pasadena. On this occasion Einstein asked him how come he had spent so much time on precision measurements of the speed of light. The answer was, "Because I enjoy it."

In the Faculty lounge of Fine Hall in Princeton University there is a marble fireplace with the following words inscribed on the mantle above it, *"Raffiniert ist der Herr Gott aber boshaft ist er nicht."* This may be translated as, "The Lord God is subtle, but he is not malicious." The story behind this inscription is as follows.

In 1921 Dayton C. Miller, a former colleague of Michelson, repeated the Michelson-Morley experiment at the top of Mount Wilson. Surprise! He got a positive result. According to him there was a definite small motion of the earth relative to the ether. He conjectured that the ether is dragged along by the earth at sea level so that the ether wind does not become noticeable until one reaches higher altitudes. If true, this result would be fatal for the Special Theory of Relativity. Einstein happened to be visiting Princeton when this result was reported and was asked to comment on the experiment. That is when he made his famous statement quoted above. The mathematician, Oswald Veblen overheard this, jotted it down, and later (1930) got permission from Einstein to inscribe this on the mantelpiece in Fine Hall.

The usual interpretation of Einstein's statement is that although it may be difficult to discover the laws of nature, it is not impossible. According to one of Einstein's collaborators, V. Bargmann, another interpretation may be possible since Einstein once said to him, "I have had second thoughts. Maybe God is malicious after all." According to Bargmann what Einstein meant was that God makes us believe we understand something when in reality we are very far from it.

Jakob Christoph Georg Joos (1894 – 1959) planned an experiment to repeat Miller's experiment on top of the Jungfraujoch (4158 m). He built the required equipment in the basement of the Zeiss factory at Jena and found no ether drift there. Although Joos was financed by the German and Swiss governments, the equipment was never transported to the top of the Jungfraujoch and Miller's results remained as a dark cloud over relativity theory long after it was accepted by almost all of the physics community.

In 1955 R. S. Shankland, S. W. McCuskey, F. C. Leone, and G. Kuerti published a detailed re-analysis of Dayton Miller's data. They found that the ether drift that he had observed was due to thermal effects in the atmosphere. Miller's results were due to systematic errors. An ether drift did not exist. Special relativity was correct.

After seeing a pre-publication version of the paper, Einstein wrote to Shankland in 1954. "I thank you very much for sending me your careful study about the Miller experiments. Those experiments, conducted with so much care, merit, of course, a very careful statistical investigation. This is more so as the existence of a not trivial positive effect would affect very deeply the fundament of theoretical physics as it is presently accepted. You have shown convincingly that the observed effect is outside the range of accidental deviations and must, therefore, have a systematic cause [having] nothing to do with 'ether wind', but with differences of temperature of the air traversed by the two light bundles which produce the bands of interference."

Chapter 5

Space Becomes Curved

"We must admit with humility that, while number is purely a product of our minds, space has a reality outside our minds, so that we cannot completely prescribe its properties a priori." Carl Friedrich Gauss, Letter to Bessel, 1830.

No scientific idea has so captured mankind's imagination and interest as Einstein's General Theory of Relativity even though there seem to be no technological developments that have emerged from this work. It is an idea so beautiful that like any great work of art it can be appreciated for its own sake. It is an idea that has totally changed our view of the universe. After Einstein's *annus mirabilis* of 1905, he was famous as one of the great scientists of the day.

Einstein's personal life, however, was not as successful as his scientific. Although one of the greatest physicists ever, he was far less successful as a husband and family man. He loved the company of women and as a young man had an out-of-wedlock daughter (Lieserl—a diminutive of Elizabeth) with his school friend, Mileva Maric. This daughter somehow disappeared shortly after she was born and to this day her disappearance and life remain a mystery. A year later, on January 6, 1903, against his family's strong opposition, he married Mileva. Their love for each other, as revealed by their letters, seems to have been more profound before they were married than after.

A little over a year, after their wedding, they had a son, Hans Albert. A second son, Eduard followed seven years later. However, their marriage was already in trouble and shortly after his move to Berlin in 1914, Einstein and Mileva separated. They divorced in 1919 when he decided to marry his cousin, Elsa Einstein Löwenthal. At that time, three years before he had been informed that he would win the prize, Einstein already specified that Mileva should receive his Nobel Prize money. The telegram, informing him that he had been awarded the Nobel Prize for Physics for 1921 was sent to him on November 10, 1922. In awarding him the Prize, the Nobel committee cited his work on the photoelectric effect, but did not mention his work on relativity or gravitation. Later in life when Einstein listed his achievements he omitted the photoelectric effect.

Regarding marriage he said that he married his cousin mainly for "the convenience of having his shirts done at home". His comment on marriage in general was, "Marriage is like a headache, something one has to endure." The amazing thing is that in spite of these personal difficulties he created some of the most profound theories of physics.

In creating the special theory of relativity, Einstein had recognized the irrelevance of uniform relative motion. It was already implicit in Newton's laws that uniform motion in a straight line had no physical significance of any kind; one could always imagine oneself moving with the same uniform motion without introducing any new forces, and thus wipe out this effect. The only way to detect any absolute uniform motion was relative to the ether. When Michelson and Morley had attempted to measure this effect they found no detectable motion of any kind of the earth with respect to this ether. Einstein then embodied all of these results in the special theory of relativity.

Now he went on to ask himself whether not only uniform absolute motion, but absolute motion of any kind, made sense. Was it really possible to talk of absolute acceleration or was only relative acceleration meaningful? He concluded that only relative motions were meaningful at all. How was it that Einstein could arrive at this conclusion when Newton's laws required forces to produce accelerations?

One of the earliest and most easily performed gravitational experiments is to simultaneously drop two objects of different mass and watch them fall. Galileo Galilei (1564 – 1642) supposedly performed this experiment by dropping different masses from the tower of Pisa and found that they landed at the same time. They fell at the same rate.

Galileo also endorsed the idea that the earth revolves around the sun instead of the orthodox idea that the earth is the center of the universe and everything else revolves around it. This brilliant scientist thus seems to have deliberately offended the clerical authorities by refusing to say that a heliocentric universe was only a possible theory, not a fact. As a consequence he was confined to his home. His refusal to acknowledge the supremacy of revealed truth over scientific facts is summarized in his statement, "I do not feel obliged to believe that the same God who has endowed us with sense, reason, and intellect has intended us to forgo their use."

In the twentieth century, astronauts, on the moon's surface surrounded by vacuum, repeated Galileo's experiment about falling objects by dropping a hammer and feather at the same time. They fell at the same rate. This result, depending on one's background, can be unexpected. Neglecting the friction of air, all objects, regardless of their mass, fall at the same rate. This fact is incorporated in Newton's law of gravitation, but is not explained by it. The most precise measurements of this equal acceleration of gravity were made first by Baron Roland von Eötvös (1848 – 1919) of Hungary in 1891 and repeated (with far greater precision: one part in 10^{11}) in 1963 by P. G. Roll, R. Krotkov, and R. H. Dicke at Princeton University.

Einstein saw this equal gravitational acceleration of all objects as a very profound fact of nature and raised it to a fundamental principle: the Principle of Equivalence. With his sure physical intuition he now extended this principle beyond the simple observed facts. He said (my paraphrase), "Consider an elevator falling freely near the surface of the earth. It is then impossible for a physicist in this elevator to perform any experiment (short of looking outside) which will allow him to decide whether the elevator is freely falling or is located in free space far from any massive bodies." Incidentally, the person in the elevator doesn't have to be a physicist.

This thought experiment has immediate consequences. One of the more surprising is that one of the basic assumptions of Special Relativity, that the speed of light in a vacuum is constant, must be given up. Instead we find that light, like matter, experiences a gravitational attraction. This is an immediate consequence of the elevator thought-experiment and can be seen as follows.

Shine a beam of light "horizontally" across the elevator. If we are located in free space, the beam will travel in a straight line to the other wall. The light beam can be made visible by placing a cloud of chalk dust in its way. If we are in a falling elevator, then the sides of the elevator will fall "down" relative to the beam and the light beam's path should appear curved upwards. Thus, such an experiment would allow us to distinguish whether the elevator is freely falling or located in free space. In order to retain the indistinguishability of these two situations, Einstein concluded that a light beam is also deflected downwards by a gravitational field and still appears to travel in a straight line when viewed by the physicist in the elevator. He reached this conclusion in 1907. The amount of light deflection however, was so small that he could think of no way of measuring it. For a very large elevator, say 100 meters across, the deflection was only about 10^{-10} cm or about 1/100 the size of an atom. By 1911 Einstein had found a way to measure this tiny effect.

He calculated that if a star emitted light that just grazed the surface of the sun on the way to us, it would be deflected by about 0.83 seconds of arc. This angle is about the same as the angle subtended by the thickness of a pencil at a distance of one mile. When Einstein wrote to George Ellery Hale, the director of the Mount Wilson Observatory to ask if such a measurement were feasible he received a positive reply. The detection of the deflection by such an angle, small as it is, was technically feasible, but only during an eclipse of the sun.

Actually, the same result had been published more than a century earlier and forgotten. In 1804 a German mathematician and astronomer, Johann Soldner (1777 – 1833) wrote in the *Berliner Astronomisches Jahrbuch auf das Jahr 1804* an article in which he calculated that a light ray passing close to the solar limb would be deflected by 0.87 seconds of arc. His calculation was based completely on Newtonian physics. The discrepancy between Soldner's 0.87 and Einstein's 0.83 seconds of arc is due to a simple arithmetical error by Einstein.

A solar eclipse was due to occur in Russia in 1914 and an expedition to test Einstein's effect was planned. World War One, however, erupted and intervened to prevent this experiment from being performed. This turned out to be fortunate for physics. The reason is that the computation using only the principle of equivalence is wrong by a factor of two. Thus, instead of 0.87 seconds of arc the deflection should really have been $2 \times 0.87 = 1.74$ seconds. Had the measurement been carried out in 1914, Einstein might have been discouraged and not gone on to obtain finally, in 1916, his field equations which predict the correct amount of deflection.

In his 1907 paper Einstein predicted another consequence of the principle of equivalence or indistinguishability of the two elevators. He concluded that in a gravitational field, light should experience a frequency shift towards the lower or red end of the spectrum. This is known as the gravitational red shift and was never directly observed during Einstein's lifetime. In 1960, only five years after Einstein's death, two Harvard physicists Robert Vivian Pound (1919 – 2010) and Glen Anderson Rebka Jr. (1931 –) succeeded in performing this experiment using the recently discovered Mössbauer effect. Einstein's prediction was beautifully confirmed.

The Mössbauer effect or Recoilless Nuclear Resonance Absorption was discovered in 1957 by Rudolf Ludwig Mössbauer (1929 –) while he was still a graduate student. This discovery earned him together with Robert Hofstadter the 1961 Nobel Prize in physics "for his researches concerning the resonance absorption of gamma radiation and his discovery in this connection of the effect which bears his name".

In general when a photon is emitted from an atom, or in this case a nucleus, the atom, or nucleus, recoils because the photon comes out with some momentum. The bigger the recoil, the more of the photon's energy is transmitted to the atom or nucleus from which it is emitted. This means that the frequency (energy) of the photon depends on the amount of this recoil. This is called a Doppler shift. So, photons are not always emitted with the same energy and do not have a sharp frequency. Mössbauer found that under certain conditions the atom containing the nucleus that emits the radiation can be bound sufficiently tightly to a crystal so that the crystal would have to recoil as a whole. Due to the macroscopic mass of the crystal, the recoil velocity is negligibly small and the photons lose almost no energy and are therefore emitted with a very sharp frequency. This was precisely what Pound and Rebka needed in order to measure the tiny shift in frequency when a photon fell 22.5 meters in Harvard's Jefferson tower.

The reason for the long delay of work on the General Theory of Relativity between 1907 and 1911 was that during this period Einstein was very concerned with something totally different, the new concept of quanta. This is the reason he neglected his favorite brainchild until 1911 when he started to prepare himself in earnest for his greatest task. Of incalculable help at this time was his mathematician friend, Marcel Grossmann who taught him the recently developed tensor calculus. Without this tool further progress would have been impossible. They published a few papers and continued to work together until Einstein left Zürich in April, 1914 to become professor at the University of Berlin. As he left, he told his friend, "They expect this goose to lay another golden egg."

What motivated Einstein to undertake such a difficult task? A partial answer may follow from two of his statements. He told Esther Salaman, a physics student. "I want to know how God created this world. I am not interested in this or that phenomenon, in the spectrum of this or that element. I want to know his thoughts. The rest are details." Similarly he told E. Gabor, "What I am really interested in is in knowing whether God could have created the world in a different way; in other words, whether the requirement of logical simplicity admits a margin of freedom."

Once settled in Berlin, Einstein summarized the work on general relativity and outlined a research program to follow. To implement the Principle of Equivalence, nothing less was required than to be able, by choosing an appropriate reference frame, to get rid of all local gravitational effects. An example of such a frame was our freely falling elevator. Not only that, but the final theory had to incorporate and transcend special relativity. Furthermore, since gravity varies from point to point, the coordinate system had to take this into account.

When Max Planck learned of Einstein's bold undertaking to find a new theory of gravity he told him, "As an older friend, I must advise you against it, for, in the first place you will not succeed, and even if you succeed, no one will believe you."

Einstein clearly saw what was needed. One could either have Newton's gravitational field together with Euclid's space and time, as Euclid had described them almost two millennia earlier, or else no gravitational fields but a curved space-time manifold. Thus, the gravitational attraction of the sun was to be replaced by a curvature of space due to the presence of the sun. Also instead of the planets being kept in their orbits by a force from the sun pulling them, they should simply follow the "shortest" path in this curved space. The mathematics for this type of geometry had already been developed much earlier by Georg Friedrich Bernhard Riemann (1826 – 1866) and was awaiting application.

A rather forced analogy of what Einstein needed to do is to consider a marble rotating about a point to which it is attached by a string. The string pulls it in and keeps it going in a circle. The same effect can be obtained if the marble rolls inside a bowl about the center of the bowl. The curved "space" of the bowl's walls keeps the bead going about the center rather than in a straight line. To stretch the analogy even further, space might be thought of as a thin elastic sheet. The attracting body's mass when placed on this elastic sheet deforms it and produces the bowl.

Why should physicists, or anyone else, have accepted such an outlandish notion? In fact Einstein himself when asked about the possibilities said, "If I am proven wrong the French will say that I am a German and the Germans will say that I am a Jew. If I am right, the Germans will declare me a citizen of Germany and the French will declare me a citizen of the world." There were many physicists that initially did not accept this theory, but soon there were in addition to the beautiful simplicity of the concepts involved two very good experimental reasons.

Walther Nernst, Albert Einstein, Max Planck, Robert Millikan, and
Max von Laue.

In May 1919, off the west-coast of Africa on the little Portuguese island
of Principe, a total solar eclipse occurred. Sir Arthur Eddington (1882 –
1944) was there to observe the eclipse and measure the deflection of the light
from a distant star whose light just grazed the sun. He had his equipment
set up well in advance of the eclipse which was due at two o'clock on May
29. The weather did not look promising. That morning there was heavy
rain. Eddington recorded what happened.

"The rain stopped about noon and about 1:30 ... we began to get a
glimpse of the sun. We had to carry out our photographs in faith. I did
not see the eclipse, being too busy changing plates, except for one glance to
make sure that it had begun and another half-way through to see how much
cloud there was. We took sixteen photographs. They are all good of the
sun, showing a very remarkable prominence; but the cloud has interfered
with the star images. The last few photographs show a few images which I
hope will give us what we need ... "

By June 3 he had analyzed some of the plates which, due to the clouds, were of poor quality. He recorded in his notebook, "...one plate I measured gave a result agreeing with Einstein." Later he wrote a poem that seems to have been inspired by poem LXIII of *The Rubaiyat* of Omar Khayyam.

> Oh leave the Wise our measures to collate.
> One thing at least is certain, light has weight.
> One thing is certain and the rest debate.
> Light rays, when near the Sun, do not go straight.

Eddington's measurements verified Einstein's prediction of the deflection of light to an accuracy of about 20%. This was sufficient to distinguish between the Newtonian prediction of 0.87 seconds of arc and Einstein's value of 1.74 seconds of arc.

The day after that historic experiment a young physicist, Ilse Rosenthal-Schneider was in Einstein's office. He was leaning against the window looking out, then turned to her, pulled out a telegram and handed it to her, "This may interest you." It was the telegram from Eddington announcing the positive result of the light deflection experiment. The young lady told Einstein that she thought that he must be just thrilled. His reply was, "You're surprised? I knew that it had to be so."

"But what would you have done if the result had been different?"

"Then I would have had to feel sorry for dear God. The theory is correct."

It was quite common for Einstein, who did not participate in organized religion, to refer to God. Thus, he stated, "God is inexorable in offering His gifts. He only gave me the stubbornness of a mule. No! He also gave me a keen sense of smell." Also, on another occasion, "Science without religion is lame; religion without science is blind." He also said, "As long as you pray to God and ask him for something you are not a religious man." and, "God does not care about our mathematical difficulties. He integrates empirically."

Even before the light deflection experiment verified this prediction, the theory had achieved an even more impressive success at its birth in 1916. Of all the planets, Mercury is closest to the sun and thus feels the effect of the sun's gravitation the most. As Johannes Kepler (1571 – 1630) had already shown and later Newton's law of gravitation had predicted, all planets move in elliptical orbits. The orbit of Mercury, however, has a prominent anomalous feature: the major axis of its ellipse rotates slowly in the direction of the motion of Mercury. This rotation returns the ellipse to its original position roughly every 23, 000 years. This is usually stated in another way: The perihelion (point of closest approach to the sun) of Mercury advances 5, 600 seconds of arc per century. Most of these 5, 600 seconds of arc per century are explained by Newtonian gravitation as due to the perturbing effect of the other planets. There remains, however, a tiny discrepancy of 43 seconds of arc per century. These residual 43 seconds of arc per century are the anomaly in Mercury's orbit and had defied explanation for over two-hundred years.

Now the field equations that Einstein had obtained, predicted orbital motions for the planets of our solar system, almost indistinguishable from the motions predicted by Newtonian mechanics except in the case of the planet closest to the sun: Mercury. For this recalcitrant planet, Einstein's theory predicted an additional advance of the perihelion of Mercury of 43 seconds of arc per century, precisely what was needed to remove this two-century old anomaly. There was no way to fudge this number in Einstein's theory, not even a little bit. This must be acknowledged as one of the greatest triumphs of the human intellect.

In the Special Theory of Relativity, Einstein had eliminated the ether. Now, in the General Theory he had shown that space has definite properties. In 1920 in a talk he presented in Leiden he pointed this out. "According to the General Theory of Relativity space is endowed with physical properties; in this sense therefore there exists an ether. As a consequence space without an ether is inconceivable. ... This ether, however, cannot be thought of as possessing the characteristic properties of a ponderable material." So, clearly this is not the ether of the earlier physicists. In modern quantum field theory, we find that space has to have even more physical properties, such as an index of refraction. Although not called by its name, the ether is back.

After Einstein's theory was published, Hermann Weyl (1885 – 1955) developed a very elegant generalization that included electromagnetic theory. He published his results in a book, *Space Time Matter*. Einstein admired the theory, but shortly after decided that it could not correspond to what actually occurred in nature. At first he called it a "master symphony", but later disagreed with it and stated, "He who has the most powerful opponent, excels." On April 19, 1918 he wrote to Weyl, "As beautiful as your concept is, I am compelled to frankly say that in my opinion it is impossible that the theory corresponds to nature."

Weyl responded on May 19, 1918. "I know only too well in an how much closer relationship to reality you are than I. ... Should you remain correct regarding the real world, then I regret to have to accuse the dear Lord of a mathematical triviality." Perhaps Weyl was not as upset as he might have been because he later confessed, "My work has always tried to unite the true with the beautiful and when I had to choose one or the other, I usually chose the beautiful."

Regarding General Relativity, Weyl also said, "It is as if a wall which separated us from the truth has collapsed. Wider expanses and greater depths are now exposed to the searching eye of knowledge, regions of which we had not even a pre-sentiment."

The General Theory of Relativity immediately grabbed the public's attention like no other scientific theory had done before. All sorts of myths arose. One of these was that there were only about a dozen people in the world who could understand this theory. Einstein tried to counter this and to explain that General Relativity followed the thinking of previous physics theories. Thus, in an interview with the *New York Times* in 1921 he stated, "There has been a false opinion widely spread among the general public that the theory of relativity is to be taken as differing radically from the previous developments in physics."

A few weeks before Einstein published his General Theory, the great mathematician, David Hilbert (1862 – 1943) at Göttingen, had published the same field equations. This came about as follows. Einstein had almost completed his work, but was having difficulties making the field equations mathematically consistent. At this time he came to Göttingen and gave a seminar in which he presented his incomplete results. At the end he pointed out his difficulties. Hilbert, who was in the audience, suggested that if Einstein used a variational principle, the resulting field equations would be automatically consistent. Einstein either did not understand what Hilbert meant or else did not want to use such an approach. Apparently, Hilbert went ahead and used just such a variational principle and published Einstein's equations before Einstein. In the original manuscript submitted by Hilbert, he had obtained the wrong field equations. Before his paper appeared, Hilbert got to see Einstein's final results and had time to make the appropriate changes in the proof copies so that the correct equations got printed.

Einstein very much valued his privacy. His wife, Elsa related the following story. While on a train trip, a man in the same compartment continued to stare and stare at him and finally asked what his profession was. An irritated Einstein answered, "I am an artist's model."

Years later, when Einstein was already a legend, an entrepreneur, who wanted his endorsement brought some cubes which formed a four-dimensional Tic Tac Toe to Einstein at the Institute for Advanced Studies. Einstein was not impressed with this individual and said. "That's not good; they are in Euclidean space."

That even someone of Einstein's stature could make silly pronouncements is evident from his statement, "There is not the slightest indication that energy will ever be obtainable from the atom."

As he grew older, Einstein like most physicists became more philosophical. His remarks reflect the change "The whole of science is nothing more than a refinement of everyday thinking." He also quoted Sir Humphrey Davy (1778 – 1829) when he stated, "One thing I have learned in a long life: that all our science measured against reality, is primitive and childlike — and yet it is the most precious thing we have." On the other hand, he was not at all enamored with philosophers such as Emmanuel Kant. "I am convinced that the philosophers had a harmful effect upon the progress of scientific thinking in removing certain fundamental concepts from the domain of empiricism, where they are under control, to the intangible heights of the *a priori*—the universe of ideas is just as little independent of the nature of our experiences as our clothes are of the form of the human body." And again, "...the physicist cannot simply surrender to the philosopher the critical contemplation of the theoretical foundations; for, he himself knows best, and feels more surely where the shoe pinches. ...Physical conceptions are free creations of the human mind and are not, however it may seem, uniquely determined by the external world."

Here are some more of his philosophical musings.

"Concern for man himself must always constitute the chief objects of all technological effort."

Near the end of his life when he was chatting with Robert Oppenheimer, Einstein commented on how his work at the early age of twenty-six had changed his life. "When it has once been given you to do something rather reasonable, forever after your work and life are a little strange."

Since General Relativity deals with space and time, physicists have been known to use this as an excuse. When Christian Møller (1904 – 1980) was giving a one-hour lecture on General Relativity he did just that. After more than an hour and a half he had filled all the boards. Pointing to these he said, "What we are facing here is a problem of space and time."

I'll let Sommerfeld have the final comment on General Relativity. "With profundity of thought and logicality of philosophical reasoning never before known in the mind of a natural scientist, and with mathematical power reminiscent of Gauss and Riemann, Einstein raised in the course of ten years an edifice before which those who follow with rapt attention his labours from year to year must stand in amazement and stupor."

Chapter 6

Kilns and Quanta

"It is called blackbody radiation. Black, because the hole in the furnace that we look at is black when the temperature is zero." R. Feynman, *Lectures on Physics.*

High in the Caucasus, chained to a rock, until rescued by Hercules, lay the titan Prometheus. A vulture pecked at his liver to maintain and increase his agony. That is how Greek legend portrayed the punishment of the giver of fire. The gods were so jealous of the power of fire that they devised this torture for the immortal that gave man this gift. The legend shows that the Greeks understood well the might of fire, so well in fact, that they viewed fire as a gift stolen from the gods. Mankind has never lost this fascination with fire. From our brutish first ancestors, who must have approached the dancing flames with reverence and trepidation, to modern young lovers cuddling in front of a fireplace, fire has retained its magic.

As a boy I spent many entranced hours in front of a blacksmith shop. Even today, more than half a century later, I can still smell the singed hooves and hear the hammer clanging on the anvil. Most vivid of all these images, however, is the picture of a black piece of iron stuck into the blacksmith's forge and a sweaty, blackened arm pulling on the shaft driving the bellows. The bellows sigh softly, flames flare a bright yellow and orange and the air shimmers above them. Lower down there are occasional blues, but mostly there are undulating reds with the white ashes showing through.

The iron changes slowly to a dull red, then to a brighter red, then orange, yellow and finally brilliant white. Black tongs reach in remove, the iron, and place it on the anvil. An arm, glistening with beads of sweat that reflect the flames, swings a hammer, and sparks shower the air. The color of the iron changes again, this time in reverse order: yellow, orange, bright red, dull red, and finally black. A sharp hiss follows as the iron is quenched in a trough of water and then returned to the forge for the process to be repeated.

I don't know how often I watched this transformation of a dull, black piece of iron into a glowing bright light and back. I only know that I never ceased to be entranced. It is a most fascinating phenomenon; it is a beautiful illustration of the temperature dependence of blackbody radiation; it is a manifestation of Lord Kelvin's second cloud obscuring the horizon of nineteenth century physics.

Why is it that a piece of iron, or in fact any substance that does not first burn, goes through these color changes when it is heated or cooled? Why is it that all materials when heated to the same temperature glow with the same color? Perhaps earlier physicists were too much struck with the beauty of these phenomena to ask such questions. Perhaps the questions were too difficult and therefore not viewed as relevant. At any rate, these questions thrust directly at the heart of the difficulty with blackbody radiation. The answer to these questions changed forever our perception of the microscopic world of the atom and opened the doors to modern solid state technology with all its consequent implications.

Before these questions could even be asked, it was necessary to realize that all materials behaved this way. One of the earliest recorded observations of this fact is due to Josiah Wedgwood (1730 – 1795), the famous maker of ceramics. He noted that he could tell whether his kilns had reached the right temperature for firing just by the color of the inside of the kiln. The color did not depend on how full the kiln was or on what he had put into it.

As so often happens, physicists did not arrive at these questions by the most direct route. They first made a detour to the stars.

Kirchhoff's name is familiar to every student of physics and engineering due to his famous Kirchhoff's laws for electrical circuits. He published these in 1845 while he was still a student. These laws apply to electrical circuits with any number of loops. Once formulated they seem almost obvious, especially so if you think of electric currents in analogy with water flow in pipes and electric voltage as the differences in height for water flow. The current law says that current cannot pile up at any point in a circuit. This means that where several wires come together, the net current into this point must be zero: the same amount of current must flow out as in. Where several water pipes come together the same amount of water must flow out as in. The voltage law says that voltage cannot pile up at any point in a closed circuit. Thus, if you go around a closed loop in a circuit and add up all the voltage drops (both positive and negative) the net result when you get back to the starting point must be zero. For the water analogy this means that if you add up all the changes in height both positive and negative in following any path of pipes that eventually leads back to where you started, the net change in altitude must be zero.

During the latter part of the nineteenth century Gustav R. Kirchhoff, was studying the spectra of stars. He used the prism spectrometer—a triangular piece of glass used to separate the various colors of light—that he and Robert von Bunsen (1811 – 1899) had recently invented. When light passes through the prism, the various colors in the light beam are bent by different amounts and separated. Since every element produces a distinct set of colored lines, they used this device to identify several elements in the sun and discover two new ones: Cesium and Rubidium. Kirchhoff also discovered some dark lines in the sun's spectrum. These, he conjectured, occurred when light of that particular color was absorbed as it passed through the sun's atmosphere. In this manner he detected helium, named after helios the Greek word for the sun. (Sadly, observing the sun's atmosphere ruined his eyesight.)

During this study Kirchhoff conducted a routine investigation of the emission and absorption of light by material bodies to calibrate the observations from the stars. In other words, to be able to compare accurately different stellar spectra he carefully measured the colors and intensities of the light emitted by material bodies at a given temperature. He did precisely what generations of boys had done with their eyes while watching the iron heat in the blacksmith's forge. Only he did it with instruments and with great precision. These measurements turned out to be seminal.

Kirchhoff found some variation in brightness among different materials and so for convenience, defined an ideal material, the blackbody, as any body which absorbs all the radiation incident upon it. The reason for the name is that any dull black surface is a good approximation of a blackbody. Next, using thermodynamics, Kirchhoff proved that the quality (color) and intensity (brightness) of radiation inside any material cavity at a given temperature is the same as that emitted by a blackbody at the same temperature. He had partly answered both our questions. What remained to answer was how the radiation depended on the temperature.

Whenever scientists find a fact that holds true for very many, or even better for all, substances they get excited. Such was the case with blackbody radiation. Here was something that was universally true; it applied to all substances. The type of radiation emitted by a blackbody became a physics problem of foremost importance. To find the mathematical function that described the intensity of radiation for every frequency inside a blackbody cavity at a given temperature became the research problem of the day, but only for those physicists concerned with the most fundamental questions. For all others it was physics as usual.

But how did the radiation depend on temperature? Several prominent physicists, both experimental and theoretical, were soon engaged in a full-scale attack on this problem. Notable among these was Wilhelm Wien (1864 – 1928). His full name was Wilhelm Carl Werner Otto Fritz Franz Wien; his colleagues called him Willi. He found that an enclosed surface with a tiny hole to peek inside provided an almost perfect blackbody. Wien succeeded in proving a general result concerning the radiation from such a blackbody cavity and derived an example of a function which agreed well with experiment at high frequency and low temperatures but, as later discovered, failed for low frequencies and high temperatures. This function is known as the Wien radiation formula.

Like many other famous German physicists of that period, Wien studied for some time (1883 – 1885) with Hermann von Helmholtz. As the son of a landowner he was expected to become a farmer, and indeed his studies were interrupted, except for one semester, from 1886 until 1890 when his father became ill and he helped to manage his lands. After the sale of his father's estate he rejoined Helmholtz and remained with him until 1896 when he assumed the chair of Professor of Physics at Aachen (Aix-la-Chapelle). From Aachen he moved to Giessen and in 1900 succeeded Röntgen to the chair in physics in Würzburg. His peregrinations ended with his move to Munich in 1920.

His two main achievements that interest us were both concerned with blackbody radiation. In 1893 he published what came to be known as the Wien Displacement Law. This law states that the temperature of a radiating body multiplied by the wavelength at which the radiation is most intense, is a constant. Using this law it is possible to measure the temperature of the sun, or a star or any other radiating body.

A year later he announced the Wien Radiation Law. It was a formula that was based on firm thermodynamic considerations and described the spectrum of blackbody radiation. This work led Max von Laue to write in Wien's obituary that "his immortal glory [was that he] led us to the very gates of quantum physics." The gates to quantum physics were crossed by

Max Planck.

In 1911, Wien received the Nobel prize in physics "for his discoveries regarding the laws governing the radiation of heat".

Wien's law, although based on firmly established thermodynamic principles and seeming to fit experimental data well, was not the final word. Detailed measurements by Otto Richard Lummer (1860 – 1925) and Ernst Pringsheim (1859 – 1917) verified the Wien formula for high frequencies, but found some discrepancies at low frequencies.

On the other hand, John Strutt, Third Baron Rayleigh of Terling Place (1842 – 1919) also derived in 1900 a completely different formula for the quality of the radiation. He combined a result from statistical mechanics (the Equipartition Theorem) with a calculation of the distribution of standing electromagnetic waves in a cubical cavity. The result, known as the Rayleigh-Jeans formula (because in 1905 James Jeans corrected an error of a factor of eight) differed drastically from Wien's formula. In fact, the Rayleigh-Jeans formula works well for low frequencies and high temperatures (where the Wien formula fails) and fails utterly for high frequencies and low temperatures (where the Wien formula works). Even worse, the Rayleigh-Jeans formula predicts the nonsensical result that any blackbody cavity will have ever more intense radiation at higher frequencies, so that if you were to peek into an ordinary stove even without a fire burning inside, your eyes would be seared with ultraviolet radiation, X-rays, gamma rays and even harder radiation. The high frequency end of the visible spectrum where all this disastrous stuff is supposed to start happening is the ultraviolet. No wonder that physicists dubbed this failure of the theory at the high frequency end, the Ultraviolet Catastrophe.

The equipartition theorem states that the average energy of a system in equilibrium is divided equally among all the possible motions of the system. This average energy is proportional to the temperature. So, to calculate the blackbody energy, Lord Rayleigh only had to enumerate the various ways in which the electromagnetic radiation in a blackbody cavity could vibrate. This was a fairly standard calculation and so seemed free of controversy.

As an aside, the equipartition theorem has an interesting history. In 1845 John James Waterston (1811 – 1883), a Naval Instructor to the East India Company's cadets at Bombay, India submitted a paper on kinetic theory, including the equipartition theorem, to the Royal Society of London. Two of the members reviewing the paper disagreed with it and thus the paper was rejected and buried in the files of the Society. Waterston had not kept a copy of his work and thus did not submit it to another journal.

He did however, in 1858, publish a paper that showed that the speed of sound could be calculated from his kinetic theory. This paper referred to his unpublished manuscript in the Royal Society Archives. In 1891 Lord Rayleigh, while writing his books on the *Theory of Sound*, searched the literature and came across this paper. At the time he was Secretary of the Royal Society, so he retrieved the unpublished manuscript and after reading it had it published with the following introduction written by him.

"The history of this paper suggests that highly speculative investigations, especially by an unknown author, are best brought before the world through some other channel than a scientific society, which naturally hesitates to admit into its printed record matter of uncertain value. Perhaps one can go further and say that a young author who believes himself capable of great things would usually do well to secure the favorable recognition of the scientific world by work whose scope is limited, and whose value is easily judged, before embarking on greater flights." So much for scientific objectivity.

John William Strutt was the oldest of seven children born to John James Strutt the 2nd Baron Rayleigh. The title harked back to John William's grandmother Charlotte Strutt who accepted the title because her husband had not wished to resign as MP and accordingly declined the honor. John William was slow to learn to speak, but once started was full of questions such as the one mentioned by his aunt Emily. "What becomes of the water spilt on the tablecloth after it has dried up?"

At the age of ten he caught smallpox and whooping cough. His parents worried about the child and considered him too frail for school. Accordingly, he was taught by a private tutor. In 1861 he entered Trinity College Cambridge and graduated with numerous honors in 1865; he was Senior Wrangler in the Mathematical Tripos. The next year he became a fellow of Trinity College. During his fellowship his travelled all over the world. On his visit to the USA he even met President Andrew Johnson.

In 1871 he married Evelyn Balfour, the sister of the future Prime Minister Arthur Balfour. However, shortly after the marriage he suffered a severe attack of rheumatic fever. The doctors suggested a trip to Egypt to restore his health. So, the young couple followed this advice and cruised up the Nile in a houseboat. During this trip he commenced work on his monumental *Theory of Sound*.

When William Strutt's father died in 1873 he became the 3rd Baron Rayleigh and took over the running of the family estates. He set up a laboratory at his residence at Terling Place and did much of his experimental work there, even after 1879 when he succeeded Maxwell as the second Cavendish Professor. He resigned this position in 1884.

The Rayleigh-Jeans formula was not Lord Rayleigh's most scholarly work; he probably considered it one of his lesser achievements. His two volume *Theory of Sound* was a masterpiece of scholarship and clear writing. These books were a major achievement and educated, as well as influenced, future generations of physicists.

Lord Rayleigh also was an able and extremely careful experimenter and succeeded in isolating and identifying the noble gas, argon. The name's origin is Greek and means "inactive" since argon is chemically inert. His discovery of this element was due to his care and persistence. He had noticed that the density of nitrogen prepared from ammonia was about half a percent lower than the density of nitrogen prepared from air. This discrepancy had to be cleared up and Lord Rayleigh worked diligently until he solved the puzzle and discovered argon as the culprit. As a consequence of this work, he received the 1904 Nobel Prize in physics "for his investigations of the densities of the most important gases and for his discovery of argon in connection with these studies".

Around 1894, the by now middle-aged, Max Planck became interested in the problem of blackbody radiation. He was well equipped to attack it since he was a specialist in thermodynamics. In fact, he had worked on the same sort of problems as Gibbs at Yale and had also succeeded in a clearer understanding of that field. It was his support of Helmholtz' theories, against those of professors at Göttingen in a prize essay, awarded by Göttingen that lost him first prize, but brought him to the attention of Helmholtz and to the University of Berlin where, at the very early age of thirty, he received the appointment as extraordinary[1] (associate) professor at the University of Berlin. This was extremely young for such a position.

While he was still quite new there he forgot which lecture hall he had been assigned to lecture in. Accordingly he stopped at the office to ask in which room Professor Planck was scheduled to lecture. The elderly man in charge of the office looked at the slim young man in front of him and said in a kind voice, "I advise you not to go there, young man. You are much too young to understand the lectures of our learned professor Planck."

There was no doubt that Max Planck was destined for an academic career: his father, grandfather and great-grandfather were all professors. At that time professors were civil servants in Prussia and young Max was raised

---

[1]literally extra to the full or ordinary

with the ideals of scholarship, honesty, fairness, generosity, and service to the state. He retained these ideals for the rest of his life.

After his father became professor in Munich, the family moved there and Max attended the famous Maximilian Gymnasium. The gymnasium is roughly the equivalent of high school, although students attend it for eight years and usually enter it after grades four or five. As a student, Max was above average, but not outstanding. After graduation in 1874, he entered the University of Munich. Here he was told for the first time that physics is essentially a complete science. In spite of this he went on to the University of Berlin in 1875 to study with Kirchhoff who again tried to discourage him from studying a subject with no future discoveries to be made. Planck persisted in physics because as he put it, "The outside world is something independent from man, something absolute, and the quest for the laws which apply to this absolute appeared to me as the most sublime scientific pursuit in life."

Planck started out as an empiricist and did not even believe in the existence of atoms. Instead he worked in his chosen field of thermodynamics. In his own words, "As the significance of the concept of entropy had not yet come to be fully appreciated, nobody paid any attention to the method adopted by me, and I could work out my calculations completely at my leisure with absolute thoroughness without fear of interference or competition."

As late as 1879, in his doctoral thesis, he had written that the concept of atoms led to contradictions. He repeated in 1882, "Despite the great success that atomic theory has so far enjoyed, ultimately it will have to be abandoned in favor of the assumption of continuous matter." By 1887, however, he had changed his mind and stated about Avogadro's hypothesis, that it must be included among "those laws ... that seem to us the most certain foundation of theoretical inquiry". This also led him to disagree with the rigid positivism of Ernst Mach. "As far as Mach is concerned, I must say that although I otherwise much appreciate the independence and sharpness of his judgment, I do not think him competent as far as the second law is concerned."

Mach denied any permanence to theories. "Theories are like dry leaves which fall away when they have long ceased to be the lungs of the tree of science." Mach's criticism did not only extend to the concept of atoms; regarding Freud's psychoanalysis he wrote. "These people want to use the vagina as a telescope through which to view the world. That is not its natural function. It's too narrow."

Once Planck had accepted atoms, he became a strong advocate. So, it is not surprising that he attacked Wilhelm Ostwald's Energetics that tried to avoid the use of atoms altogether. "I consider it my duty to warn most emphatically against the further development of Energetics in the direction it has recently taken, which signifies a serious backward step from the current results of theoretical work and can only have the consequences of encouraging young scientists in dilettantish speculations instead of in a thorough grounding in the study of established masterpieces."

As a teacher Planck gave the following advice to his student, Max Theodor Felix von Laue (1879 – 1960), "My maxim is always this: consider every step carefully in advance, but then if you believe you can take responsibility for it, let nothing stop you." He also believed in the unity of all science. "For science, objectively considered, forms an internally closed whole. Its division into fields is not based on the nature of things."

To return to blackbody radiation and recapitulate. The problem Planck attacked was the following. There existed in nature this supposedly universal function describing the quality and intensity of radiation by a blackbody. This function had to satisfy the general conditions derived by Wien. Furthermore, this function coincided with the Rayleigh-Jeans formula for low frequencies and with the Wien formula for high frequencies, but failed for both formulas at one or the other end. (Planck was probably unaware of Lord Rayleigh's result at the time, since it had been published only a few months earlier.) Planck was very much aware of the failure of the Wien formula at low frequencies and in a letter to Wien criticized James Jeans for not accepting the strong experimental data that showed discrepancies with the Wien law. "He is the very model of a theorist as he should not be, just as Hegel was in philosophy: if the facts don't fit, so much the worse for them."

The first thing Planck did was to restate this problem in terms of thermodynamic quantities. He then tried to derive a formula for this function and promptly wound up with Wien's formula.

I have presented the problem as seen from today's perspective. In the fall of 1900 the facts were by no means as clearly spelled out or obvious. Fortunately two experimentalists, Heinrich Rubens and Ferdinand Kurlbaum checked the Rayleigh-Jeans formula in great detail and found it correct at low frequencies. They were working in the same building as Planck and related their findings to him a few days before the meeting of the Berlin Academy at which they formally presented their results. These few days sufficed for Planck to modify his approach so that it yielded Wien's formula at high frequencies and the Rayleigh-Jeans formula at low frequencies. This new formula, now celebrated as the Planck radiation formula, was ready in time to be included as a comment to follow the report by Kurlbaum. The title of the comment was, *On an improvement of Wien's radiative formula.* Further experimental tests of this formula soon followed and showed that it agreed with experiment for all frequencies and temperatures.

Planck followed his "comment" with a paper, *On the Theory of the Energy Distribution Law in the Normal Spectrum.* Later he wrote about this paper, "I had already been struggling with the problem of the equilibrium of matter and radiation for six years without success; I knew that the problem is of fundamental significance for physics; I knew the formula that reproduces the energy distribution in the normal spectrum; a theoretical interpretation had to be found at any cost, no matter how high."

What had Planck achieved? In one sense he had only fudged a formula that agreed with Wien's formula and the Rayleigh-Jeans formula in those regions where they were valid. On the other hand, the fact that his formula was undoubtedly the correct mathematical expression, was like nature allowing him a peek at the answer in the back of the book. Planck fully realized this and knowing the correct answer now worked feverishly in an attempt to find a theoretical way to derive this formula. Every such attempt, however, led him back to the Rayleigh-Jeans law. Finally, almost as an "act of desperation", he decided to divide the energies in his calculation into little lumps of energy. At the end of the calculation he intended to let the size of these lumps go to zero so that there would be no restriction on the smallest amounts of energy possible. His initial attempt succeeded, but when he let the energy of the lumps go to zero, his formula reverted to the Rayleigh-Jeans law. Thus, against his will, he was forced to assume that energy comes in small lumps. Of course it sounds much more profound to use the Latin name for small lumps and so he called them "quanta" (the plural of "quantum").

Planck worked hectically for the next eight weeks, completed this work and many of its ramifications. On December 4, 1900 (the birthday of quantum physics) he presented his results to the German Physical Society.

In this seminal work he introduced two fundamental constants into his radiation formula: Boltzmann's constant k and what is now called Planck's constant h. He also determined their values by fitting his formula to the experimental curves. Planck always referred to h as "the quantum of action" and never as Planck's constant, as it is now known. He also named k after Boltzmann out of respect for the latter. The quantity Nk (N is Avogadro's number) equals the gas constant R, a number well known from measurements on gases. So, having determined k this simple formula gave him a way to evaluate N. Now, knowing N together with measurements from electrolysis that yielded the total charge Q required to neutralize one mole of ions, allowed Planck to calculate $e = Q/N$ the charge of J. J. Thomson's electron. Both these numbers N and e obtained directly from Blackbody Radiation measurements turned out to be in good agreement with what was then known. So by 1900 he could state, "Irreversibility leads necessarily to atomism."

This earth-shaking paper, destined to be the main force in over-throwing the deterministic Newtonian world, was not even deemed important enough by the German Physical Society to be published in full. Only an outline was published and outside Germany this paper attracted almost no attention at all. Planck himself, however, realized the importance of this work. During a walk in the forest in the suburbs of Berlin, the Grunewald, Planck indicated to his son that he felt that he had made a discovery which could be compared only to the discoveries of Newton. This assessment was indeed correct. Later he recalled, "It gave me particular pleasure, as compensation for the many disappointments I had encountered, to learn from Ludwig Boltzmann of his interest and complete agreement with my new line of reasoning."

Planck also recognized that the constants h and k as well as the speed of light c and Newton's gravitational constant G provided a system of natural units. "With their help we have the possibility of establishing units of length, time, mass, and temperature, which necessarily retain their significance for all cultures, even unearthly and nonhuman ones."

On May 28, 1927, after the full theory of quantum mechanics had already been established, Max Planck visited Amsterdam to receive the Lorentz Medal. At this time Hendrik A. Kramers (1894 – 1952) was asked to give a seminar on quantum mechanics, which Planck attended. Kramers opening sentences were typical. *"Ich fühle mich heute wie ein Kind, das seinem Vater sein Spielzeug zeigt. Ich weiss, er wird schon horchen, eben weil er der Vater ist."* (Today I feel like a child that is showing its toys to its father. I know that he will listen, just because he is the father.) On this occasion, Ehrenfest asked Planck if he did not yearn for the good old days, when physics still worked with clear-cut models and deterministic theories.

Planck responded, *"Man soll sich niemals zurücksehnen nach einer Sache von der man eingesehen hat dass sie unrichtig war."* (One should never pine for something that one has come to realize was incorrect.)

In an essay, the *Origin and Development of the Quantum Theory* he wrote, "The whole history of its development reminds me of the well-proved adage that to err is human." Interesting also is the fact that when Max Planck recommended Einstein for a chair in physics at the University of Berlin he wrote, "Although he may have shot beyond the target with his light quantum hypothesis, this should not be held against him." He was referring to Einstein's 1905 paper explaining the photoelectric effect. Not only was Einstein's explanation correct, but it also provided positive proof that Planck's quanta were necessary.

There are not many humorous anecdotes about Planck since he was a very sober, hard-working scientist of great courage who later stood up to the Nazis in his protest of their treatment of the Jews. When as president of the Prussian Academy of Sciences he visited Hitler he stated, "I want to inform you that without the Jews there is no mathematics and physics possible in Germany." This statement should be contrasted with the official Nazi dogma as expounded in Philipp Eduard Anton von Lenard's (1862 – 1947) book *German Physics*. "Jewish physics is merely an illusion—a perversion of basic Aryan physics. Science like every other human product is racial and conditioned by blood." No wonder that the selection committee, consisting, among others, of Planck, Laue, and Haber when looking for a possible replacement for Rubens had earlier rejected the two fierce anti-Semites, Johannes Stark (1874 – 1957) and Lenard with the words, "Because of their passionate and not always objective opposition to the new theoretical physics, these important scientists would endanger the fruitful collaborations of the Berlin physicists." Planck also wrote to Wien regarding Lenard. "He confuses subjective intuition with objective facts, believes that he grasps material that he does not understand, and does not recognize the limits of his importance. That is very dangerous for an academic teacher."

After Niels Bohr became famous, many physicists were invited to his institute in Copenhagen and all came, except Planck when he was invited during the Nazi regime. He explained that, "On my earlier travels I felt myself a representative of German science and was proud of it. Now, I would have to hide my face in shame."

In a letter to Bohr he expressed his moral belief. "The highest court is in the end one's own conscience and conviction—that goes for you and Einstein and every other physicist—and before any science there is first of all belief. For me, it is a belief in a complete lawfulness in everything that happens."

It took time before Planck's ideas began to be used by other physicists, but he knew they would. His attitude to new theories is illustrated by the following statement, "An important scientific innovation rarely makes its way by gradually winning over and converting its opponents: it rarely happens that Saul becomes Paul. What does happen is that its opponents gradually die out, and that the growing generation is familiarized with the ideas from the beginning." This is very similar to a statement by Hermann Ludwig Ferdinand von Helmholtz (1821 – 1894). "The originator of a new concept finds, as a rule, that it is much more difficult to find out why other people do not understand him than it was to discover the new truths."

Planck also stated, "The assumption of an absolute determinism is the foundation of every scientific inquiry."

According to Heisenberg, Planck's belief in an objective world evolving according to strict causality was irrational and this irrationality was due to Planck's "religious moral conception of life ..., which enable him to walk a straight and almost too certain a path even where immeasurable epistemological abysses threaten on either side."

Pauli felt that Heisenberg had been too kind in his criticism, perhaps because he had won the Planck medal. He felt that Planck's epistemology was "sloppy" and wrote to Heisenberg, "May the spirit that dominates Planck's scientific production and personal life not take over your publications and life too strongly." And further, "If you grant that statements about the reality of the outer world make sense, you give the devil of 'ism-philosophy' your little finger, and soon he will take the whole hand."

Planck remained cautious and conservative in his scientific thinking. Ten years after introducing the constant named after him he still wrote, "The introduction of the quantum of action h into the theory should be done as conservatively as possible, that is, alterations should only be made that have shown themselves to be absolutely necessary." This is also supported by his statement regarding scientists as they age and gain authority. He stated that they should display, "an increased caution and reticence in entering new paths". When in 1911, Walther H. Nernst (1864 – 1941) got the Belgian industrialist, Ernest Solvay (1838 – 1922) to finance the first Solvay conference to study the problem of radiation and quanta, Planck thought it was still premature.

One success of this Solvay conference was that soon after, Paul Ehrenfest (1880 – 1933) and Henri Poincaré (1854 – 1912) independently proved that the blackbody radiation law required quanta. The paper by Poincaré even convinced James Jeans who had held out for ten years against the quantum theory.

Nernst was Planck's colleague at the University of Berlin. He had formulated what is now called the Third Law of Thermodynamics, that it is impossible to reach the absolute zero of temperature. He was a very practical physicist and invented a light source called the Nernst glow lamp. Eventually he sold the patent for one million Reichsmark, (what today would exceed $10,000,000.00) just before the incandescent light bulb was invented.

For a while, after World War One, Nernst carried a gun. One evening in the Tiergarten (zoo) somebody stepped out of the bushes and asked him for the time. He was so rattled that he could not find his watch but, pulled out his gun instead. He then demanded the stranger's watch. It was rumored at the University of Berlin that thereafter Nernst had two watches.

Nernst was also famous for his *"Gold Witz"* (literally "gold joke") lecture that he repeated every May, to start off his electrochemistry lectures. At the beginning of this lecture he would casually inquire of the students if any of the gentlemen had any gold with him. Of course, students did not carry gold with them into the lecture room. So, nobody answered. He then reached into his pocket and produced some gold medals. "Here I have a few medals. This one I received in 1908 from the Royal Society for my investigations into electrochemistry. This one I received with the Nobel Prize for my Heat Theorem" (third law) and so on. Without ever stating what he wanted to do with the gold in the lecture he continued to elaborate for a whole hour about his achievements and fame.

Although Planck early on permitted women to attend his lectures, and even accepted women students, his belief at that time was, "In general it cannot be emphasized strongly enough that Nature itself has prescribed for a woman her place as a mother and housewife and that natural laws cannot be ignored under any circumstances without serious damage, which in this case would appear especially in the next generation."

By 1913 he had mellowed and was championing the rights of the outstanding woman scientist.

D. M. Bose had this to say, "After attending Planck's lectures (I knew) what a system of physics meant in which the whole subject was developed from a unitary standpoint and with the minimum of assumptions." This is in marked contrast to what Sir Rudolf Peierls stated in his lecture *The Glorious Days of Physics: My Life as a Physicist.* "..., but in any case Planck's lectures were among the worst I ever attended; he used to read the text of one of his books so that you could follow your copy, running your finger along line after line."

After the papers incorrectly reported that he had received the Nobel Prize, Planck wrote a letter to Wien. "The (Nobel business) has so far brought me only irritation because of completely nonsensical newspaper reports. I ask therefore only for your silent sympathy." In 1918 he did receive the Nobel prize "in recognition of the services he rendered to the advancement of Physics by his discovery of energy quanta".

Planck's wife Marga revealed that, "He only showed himself in all his human qualities in the family." He also confessed, "How wonderful it is to set everything aside and live entirely within the family."

Planck's life, especially his later years, was clouded by tragedy and two world wars. His wife Marie died in 1909, leaving him with two sons: Erwin and Karl and twin daughters, Margarete and Emma. In 1911 he remarried Marie's niece, Marga von Hösslin. World War One took Planck's youngest son, Karl, in 1916. Margarete died in childbirth in 1917, as did Emma in 1919. His biggest disaster, however was the death of his oldest son, Erwin. On February 23, 1945, five days after Max Planck still expected the sentence to be commuted, the Nazis hanged Erwin for his involvement in the July 1944 assassination attempt against Hitler. Max Planck was devastated by grief and lost interest in life. Today there is a small street in the district of Schwabing in Munich named "Erwin Planck Street".

Yet more disasters were Planck's lot. To escape the bombings, he had left his home in Grunewald near Berlin in 1943 and moved to the small town of Rogätz. Soon after his home and, most tragically for him, his books and notes were destroyed. But as the war drew to a close, even Rogätz was not safe and he and his wife were forced to sleep in haystacks in the fields. Planck became totally incapacitated with pain when his vertebrae fused. Robert Pohl, a professor of experimental physics at Göttingen, was able to get the American occupation troops to rescue Planck and bring him to Göttingen to live with a niece who still had a home. He spent five weeks in a hospital. After all his experiences, it is not surprising that when Planck was asked what he thought of civilization, he responded, "We should have it."

After World War Two, although already eighty-seven years old, Planck again assumed the presidency of the *Kaiser Wilhelm Gesellschaft* and helped in the reconstruction of German science as he had done after World War One.

The Kaiser Wilhelm Gesellschaft was founded in 1911 by August von Trott zu Solz, the Prussian Minister of Culture "to foster science and research". The foundation supported fundamental research from a rather large endowment fund, but almost collapsed during Germany's huge inflation following World War One. The second threat to the foundation came during the Nazi regime. Max Planck had been president of the Kaiser Wilhelm Gesellschaft from 1930 until 1936 when, in protest over the Nazi dismissal of prominent Jewish scientists, he resigned. In 1948, the name of the foundation was changed to the Max-Planck-Gesellschaft and is now the main source of funding for German research.

Sommerfeld had recognized Planck's achievement early and sent him a poem to celebrate this seminal work in quantum mechanics.

> *Der sorgsam urbar macht das neue Land,*
> *Dieweil ich hier und da ein Blumensträusschen fand.*

Translation:

> Whose care with new growth made the land abound
> While I, no more than here and there a nosegay found.

Planck ever gracious responded with a poem of his own.

> *Was ich gepflückt, was Du gepflückt.*
> *Das wollen wir verbinden,*
> *Und da sich eins zum andern schickt*
> *Den schönsten Kranz draus winden.*

Translation:

> What you did choose, what I did choose
> We want to bind as one
> And since the two harmoniously fuse
> Let the most beautiful wreath be done.

Chapter 7

The Hydrogen Atom: Plum-pudding or Planet

"I am not saying this in order to criticize, but this is sheer nonsense." N. Bohr

The glow of neon lights is a familiar sight in our cities. They tell us where to EAT or which HONEST car or furniture dealer to buy from. They are a part of the very fabric of our society. Perhaps that is why we do not really see them and marvel at their colors. Yet, it is a very interesting phenomenon, this glow of neon lights and the colors they display. Why does the gas in such a tube glow when a current passes through it? Why is the color of a given gas always the same? This is not the glow of a heated body like the glow of the filament of an incandescent light bulb; it is something totally different. Just how different the two phenomena are is most readily seen by passing the light from a light bulb and from a discharge tube through a prism. The light from the filament produces a more or less continuous spectrum like a rainbow. On closer inspection a few dark lines appear in the spectrum. On the other hand, the light from a discharge tube produces a series of very distinct bright lines of brilliantly pure colors. These colors are more characteristic of the gas in the tube than fingerprints are for an individual person.

The study of these spectra of atoms and molecules was a major occupation of a host of nineteenth-century physicists. Many spectra were catalogued and a search for systematic rules describing these myriads of colors was attempted. As chance would have it, someone from outside the field made the first important breakthrough.

One day a chemist in Switzerland visited a schoolteacher friend, Johannes Jakob Balmer (1825 – 1898), who enjoyed playing with numerological puzzles. In fact, the title of his *Habilitationsschrift* (a sort of second Ph.D. thesis required before one was allowed to teach at a university in Switzerland) was "The Prophet Ezekiel's vision of the Temple broadly described and archetectonically explained." In this thesis he "explained" and disclosed the secret numerical relationships in the Temple. He had also calculated the number of steps in pyramids.

Sometime during the evening Balmer complained to his friend that he had no interesting puzzles to work on; he didn't just want to work on artificial puzzles, but wanted something really worthwhile and difficult. The chemist had just the thing to challenge Balmer. He gave him the following four numbers: 6562.8, 4861.3, 4340.5, 4101.7. These are the wavelengths of the first four lines of the hydrogen spectrum and had been recently measured and were believed to be related. He asked Balmer to see if he could find a simple formula for them. Within days Balmer had found a very simple formula indeed. The frequencies for these lines were proportional to the differences of the reciprocals of squares of integers. In particular the frequencies corresponding to these wavelengths were proportional to

$$\frac{1}{4} - \frac{1}{9} \ , \ \frac{1}{4} - \frac{1}{16} \ , \ \frac{1}{4} - \frac{1}{25} \ , \ \frac{1}{4} - \frac{1}{36} \ .$$

Written another way these numbers look like:

$$\frac{1}{2^2} - \frac{1}{3^2} \ , \ \frac{1}{2^2} - \frac{1}{4^2} \ , \ \frac{1}{2^2} - \frac{1}{5^2} \ , \ \frac{1}{2^2} - \frac{1}{6^2} \ .$$

On the basis of this formula, which is now called the Balmer series, he predicted a completely new fifth spectral line whose frequency corresponded to $\frac{1}{2^2} - \frac{1}{7^2}$. Sure enough, this line was soon found and even a sixth line corresponding to $\frac{1}{2^2} - \frac{1}{8^2}$. So, here again nature had permitted a peek at the answer in the back of the book. The trick was to know how to use this answer.

Johann Jakob Balmer was the son of Chief Justice, Johann Balmer in the half-canton of Basel-Landschaft. The young Johann studied mathematics at the University of Karlsruhe and the University of Berlin. He returned to Basel and in 1849 earned a doctorate from the University of Basel. Balmer remained in Basel for the rest of his life as a teacher of mathematics at a high school for girls. At the age of forty-three he married and had six children.

Although his specialty was geometry and he even taught for a while at the University of Basel, he is only remembered for his formula in physics.

Balmer published his results in 1885. More general formulas, along Balmer's approach soon followed for the spectra of other atoms. In 1908

Walter Ritz (1878 – 1909), a former classmate of Einstein found that all atomic spectral lines had, just like the Balmer series, frequencies proportional to the difference of two fairly simple terms. This was the last major work this young man accomplished. He had been seriously ill from 1904 to 1906 when he again started to publish in spite of his poor health. Also in 1908 Ritz had a strong disagreement with Einstein about aspects of electrodynamics dealing with blackbody radiation. Their feud was carried out in a series of articles published in the *Physikalische Zeitschrift*. In 1909, Einstein and Ritz resolved the dispute and published a joint paper. Each one stated his position and Einstein admitted his error. Overwork and malnutrition took their toll and Ritz died shortly after. Nevertheless, his *Ritz Combination Principle* added another piece to the puzzle, which now started to assume a vague shape, and began to attract some keen minds.

How were these discrete spectral lines to be explained? Clearly one would have to use Maxwell's equations since they governed all electromagnetic phenomena, including light. In order to use these equations, however, one first had to determine how the charges were distributed and moving in an atom. All this at a time when some physicists still believed that atoms were a mere theoretical construct with no reality as a basis.

What was needed was a mathematical model of a stable atom. Now the same Lord Kelvin of "the dark clouds on the horizon of physics" fame had already made some suggestions in this direction. Sir J. J. Thomson, who discovered the electron in 1897, took up these ideas.

At twenty-eight years of age, J. J. Thomson had casually applied for the Cavendish chair vacated by Lord Rayleigh, according to his original commitment to vacate the chair after five years. To his surprise Thomson got the position. He later said, "I felt like a fisherman who with light tackle had casually cast a line in an unlikely spot and hooked a fish much too heavy for him to land. I felt the difficulty of following a man of Lord Rayleigh's eminence."

This very modest and able physicist, head of the Cavendish Laboratory in Cambridge, now commenced a systematic study of the structure of the atom. He was not motivated by a desire to understand the spectra of atoms but rather by a desire to understand the periodicity that Dimitri Ivanovich Mendeléev (1834 – 1907) had discovered in the chemical elements, what we now call the periodic table.

The most obvious candidate as a model for the atom was the so-called Saturnian Model of the Japanese physicist Hantaro Nagaoka (1865 – 1950) in which electrons were conceived as orbiting in a ring about a central nucleus. Nagaoka was the most famous Japanese physicist at the beginning of the twentieth century. He later inspired another young Japanese, Hideki Yukawa (1907 – 1981), to go into physics. As Yukawa later wrote, "Probably the one most decisive factor [was that] one could find among the Japanese ahead of one such a great physicist." Nagaoka's model however, had one very serious flaw: it was mechanically unstable. That is, the electrons would always oscillate so that the whole system would fly apart in a very short time. This difficulty did not occur in a model devised by Thomson, the Plum-pudding Model, in which many electrons flitted about like raisins in a cloud of positive charge (the pudding). Using this model, Thomson was able to provide a satisfying, albeit qualitative, explanation of the periodic table.

Both the Saturnian and Plum-pudding Model suffered from a second instability. As Maxwell's equations showed, and Hertz had verified, accelerated charges radiate. Thus, in either model, the electrons in their orbits had to radiate, lose energy and spiral into the center like spent artificial satellites crashing into the earth. Again, Thomson showed that this spiraling in would be greatly retarded if many electrons were present, since their mutual repulsion would keep them from spiraling in. The eventual decay of these atoms might also possibly explain the radioactivity of atoms recently discovered by Becquerel. Thus, the Plum-pudding Model stuffed with lots of electronic raisins seemed the most likely candidate.

One of the figures that dominated the physics of the first half of the twentieth century was Niels Bohr (1885 – 1962). Both he and his older brother Harald were destined for fame: Harald as a mathematician and Niels as a physicist. They were also both excellent soccer players, at the level of professionals. As boys, both were somewhat different from other children. According to Niels Bohr, one day on a tram their mother was reading to them and they were both slack-jawed with interest in the story. A stranger observing these two somewhat dull-looking children sympathized with their mother, "It must be very difficult for you to bring up these two children. Poor boys!"

In 1911 Niels Bohr, replete with a new Ph.D. in physics, arrived in Cambridge to work with Sir J. J. Thomson and learn more physics of the atom from him. It was an exciting time in physics. The world was still getting used to Wilhelm Röntgen's discovery in 1895 of a new type of radiation which he called X-rays.

A few months later in 1896, Antoine Henri Becquerel's (1852 – 1908) discovered radioactivity. Also, in 1906, Sir J. J. Thomson had shown that the electron was a charged particle with a fixed charge to mass ratio. Ironically, only twenty-one years later in 1927, his son Sir George Paget Thomson (1892 – 1975) would show, beyond the shadow of a doubt, that the electron is not just a particle but also has a wave nature.

More than thirty years later, G. P. Thomson recalled, "It is difficult for a young physicist to realize the state of our science in the early 1920's. It was not just that the old theories of light and mechanics had failed. On the contrary! You could say that they had succeeded in regions to which they could hardly have been expected to apply, but they succeeded erratically. And over the whole subject brooded the mysterious figure of h." The h he is referring to is Planck's constant.

Sir J. J. Thomson was a strange sort of experimentalist. He was able to diagnose experiments with uncanny ability, but as his son explained, "JJ was so clumsy that it was better to keep him away from experimental equipment and let someone else handle the apparatus." He was also very much an empiricist as this statement by him in the journal *Science* (August 27, 1909), several years after Einstein's Special Relativity, shows. "The ether is not a fantastic creation of the speculative philosopher; it is as essential to us as the air we breathe."

Unfortunately (or perhaps fortunately for physics) JJ's and Bohr's personalities did not mesh. Although JJ was very friendly and even invited Bohr over for dinner, Niels Bohr and Sir Thomson could never work together. JJ liked to work alone, whereas Bohr needed someone with whom to discuss while working. Furthermore, Thomson worked with qualitative models and with analogies whereas Bohr wanted mathematically precise concepts. As Bohr used to say, "Things needed not to be very exact for Thomson, and if it resembled a little, it was so." Thus, Bohr's criticisms of J. J. Thomson's models were certainly not well received. Finally Bohr decided to go to Manchester to take a six-week course on experimental technique, with one of JJ's former students, Ernest Rutherford (1871 – 1937). Here Bohr soon became even more interested in the structure of the atom.

Rutherford, was a huge bull of a man with a voice to match. He hailed from New Zealand. Shortly after Rutherford had arrived at Cambridge to work with JJ Thomson as the first foreign student, it was realized that he was far above the ordinary. One of his classmates wrote, "We have got here a rabbit from the antipodes, and he's burrowing mighty deep."

After studying with Sir J. J. Thomson, Rutherford accepted a chair at McGill University in Montreal. In recommending Rutherford for the position at McGill, JJ wrote as follows. "I have never had a student with more enthusiasm or ability for original research than Mr. Rutherford, and I am sure if elected, he would establish a distinguished school of physics at Montreal. I should consider any institution fortunate that secured the services of Mr. Rutherford as professor of physics." Once settled in Montreal, Rutherford started his famous scattering experiments.

Shortly after Rutherford arrived at McGill, John Cox, the chairman of his department after observing him at his research, pulled him aside and told him, "I think I had better take your classes and do the teaching work. You keep on doing what you have to do." While still at McGill, Rutherford refused a very lucrative position at Yale with the comment, "They act as if the university was made for students."

The following story illustrates how seriously experimental physics was viewed outside Rutherford's laboratory. In 1902, Rutherford and Frederick Soddy (1877 – 1956) got a Hampton liquid air plant. In most institutions this was a fad but Rutherford and Soddy used it to condense both radium and thorium and show their material nature. However once, when liquid air was needed for research, Rutherford learned that the "prepared supply had been taken to a church social for a demonstration". He was less than pleased.

Rutherford's patron at McGill was MacDonald of tobacco importing fame. Nevertheless, MacDonald thought that tobacco was a filthy habit and Rutherford often had to air out his office quickly when MacDonald was coming. In spite of the good support, money was scarce and Rutherford never forgot his experience. On one occasion, many years later when he had a large research establishment going, he exhorted his workers with the statement, "We've got no money, so we've got to think." The following story may also illustrate the frugality practiced at his lab. Sir Edward Crisp Bullard (1907 – 1980) as a Research Student in Rutherford's Cavendish Lab needed a one-inch diameter steel pipe and with great misgivings went to Lincoln the tight-fisted lab steward. This fine gentleman handed Bullard a hacksaw, pointed to an abandoned bicycle in the courtyard and told him to saw off the bar.

In 1907, Rutherford returned to England to become Langworthy Professor of Physic at Manchester. After JJ's retirement, Rutherford moved to the Cavendish Laboratory to assume his former teacher's chair. His work became so dominant that when his friend A. S. Eve once playfully depreciated Rutherford's immense contributions to physics charging that "he rode the crest of a wave", the latter responded quite honestly, "Well, I made the wave, didn't I?"

Rutherford was also very much aware of the practical aspects of doing science. One of his statements on the subject made the cover of the *Bulletin of the Institute of Physics* as an admonition to all scientists. "It is essential for men of science to take an interest in the administration of their own affairs, or else the professional civil servant will step in—and then the Lord help you."

About a pompous government official Rutherford was heard to say, "That man is like the Euclidean point, he has position without magnitude."

As stated, Rutherford was not greedy. For example when in 1920 Marie Curie was given a full gram of radium by public subscription and Sir James Chadwick (1891 – 1974) remarked to Rutherford that it was a pity no-one gave him such a large amount of the precious element Rutherford replied, "My boy, I'm glad no-one has. How would I justify each year the use of one gram of radium."

When Rutherford was doing his experiments at Manchester, he needed radium as a radioactive source. At that time he had less than 20 mg of radium. The Vienna Academy of Sciences lent him 350 mg of radium bromide in addition to 350 mg that they had sent to Sir William Ramsay (1852 – 1916) at University College London to share with Rutherford. The first 350 mg, however remained in London since Ramsay and Rutherford did not get along. Rutherford kept his 350 mg throughout World War One. After the war the British government wanted to confiscate this radium. Rutherford insisted that this radium should be purchased. He was then able to use the money to help the almost broke Vienna Institute. In this way he endeared himself to the Austrian scientists.

In addition to ruthlessly honest, Rutherford was also completely sure of his work. In reply to Sir Arthur Eddington who remarked after a dinner party that possibly electrons were only mental concepts Rutherford got up and exclaimed, "Not exist? Not exist! Why I can see them as plainly as I can see that spoon in front of me." Of course this was hyperbole of epic proportions.

Rutherford was so completely the experimental physicist that he was heard to exclaim in his laboratory, "Don't let me catch anyone talking about the Universe in my department." Thus, it is not surprising that he was less than sympathetic to some of the newest theoretical developments. Relativity was one of those theories for which he never developed an interest or understanding. In response to the writer, Stephen Leacock's question as to what he thought of Einstein's Theory of Relativity, he responded, "Oh, that stuff. We never bother with that stuff in our work." On another occasion, with a stunning lack of prescience, he stated, "The energy produced by the breaking down of the atom is a very poor kind of thing. Anyone who expects a source of power from the transformation of these atoms is talking moonshine."

Sir Ernest believed that an experiment should lead to physical conclusions that were intelligible to anyone of reasonable intelligence. He stated this in the following terms to his friend Lord Tweedsmuir, "An experiment is not complete until it has been expressed in simple and correct English." Later he amplified this. "You have not understood something unless you can explain it in terms that can be understood by an English barmaid." No one knows why he chose English barmaids. In Göttingen he expanded on this theme, "I'm always a believer in simplicity, being a simple person myself."

Rutherford also did not like experiments that required statistics to extract meaningful results. This he stated as follows, "If your experiments needs statistics, you ought to have done a better experiment." On the other hand, Piotr Leonidovich Kapitza (1894 – 1984), got to be his student by using statistics. When he applied to Rutherford for a position the latter stated that all positions were filled. At this point Kapitza asked him how accurately he could measure results in his experiments. Rutherford replied, "To 10%." Kapitza pointed out that since there were already thirty students in Rutherford's department one more student was well within the experimental accuracy. Another version states that Rutherford replied, "Three per cent." This was barely enough. In any case, Kapitza was to become one of Rutherford's most famous students and do much of the crucial experimental work on liquid Helium.

A couple of years after Rutherford's death Kapitza lamented that event, as well as the enforced secrecy that came with World War Two. "The year that Rutherford died there disappeared forever the happy days of free scientific work which gave us such delight in our youth. Science has become a productive force. She has become richer but she has become enslaved and part of her is veiled in secrecy. I do not know whether Rutherford would continue to joke and laugh as he used to."

Perhaps Rutherford's dislike for statistical or probabilistic arguments went back to having to study together with his students to learn the requisite mathematics to be able to analyze his most famous experiment. Another version states that apparently Rutherford's mathematical abilities were rather limited and his famous scattering formula was derived by a young mathematician, R. H. Fowler.

Rutherford and Hans Geiger (1882 – 1947) had been studying the scattering of alpha particles by shooting them through thin metal foils. Einstein had this to say about these experiments, "It is like shooting birds in the dark in a country where there are not many birds." The results had not proved particularly exciting and Rutherford assigned a young student, Ernest Marsden (1888 – 1970) to do a more careful experiment. As Marsden recollected half a century later, "He turned to me and said, 'See if you can find some effect of alpha particles directly reflected from a metal surface'. I do not think he expected any such result, but it was one of those hunches that perhaps some effect might be observed and that in any case the neighboring territory of this Tom Tiddler's ground [1] might be explored by reconnaissance." Very soon afterwards Marsden reported that a few of the alpha particles were deflected through very large angles, almost straight back. This was a most surprising result and even Rutherford was not totally sure if this was true. His reaction later was, "It was quite the most incredible event that has ever happened to me in my life. It was almost as incredible as if you had fired a fifteen-inch shell at a piece of tissue paper and it came back and hit you."

At the time, Marsden was working under the supervision of Hans Geiger who now joined him in more careful work. Again Rutherford reported, "Two or three days later Geiger came to me in great excitement and said, 'We have been able to get some of the alpha particles coming backwards'."

In the introduction to their paper, published in the *Proceedings of the Royal Society*, 1909, Marsden and Geiger start by saying that "β-particles have been shown to reflect off metal foils." They then go on to say, "For α-particles a similar effect has not previously been observed, and is perhaps not to be expected on account of the relatively small scattering which α-particles suffer in penetrating matter." In the next paragraph they then make their strong claim. "In the following experiments, however, conclusive evidence was found of the existence of a diffuse reflection of the α-particles."

This amazing result required explanation and so Rutherford wound up taking notes in a mathematics class next to his physics students. All of this happened just prior to Bohr's arrival at Manchester. Using his newly learned mathematics and the data collected by Geiger and Marsden, Rutherford succeeded in showing two important results:
1) The atom is not like a plum pudding, but is rather like a miniature solar

---

[1] Tom Tiddler's ground refers to a gold mine.

system with almost all the mass concentrated near the center and the much lighter electrons revolving around this center.

2) The number of electrons in an atom is quite small, in fact, approximately equal to one half of the mass number of the atom.

Ernest Rutherford facing John Ratcliffe under the sign "Talk Softly Please" in the Cavendish Laboratory, University of Cambridge.

This allowed the great man to state sometime early in 1911 that, "I now know what the atom looks like." In the May 1911 issue of the *Philosophical Magazine*, Rutherford announced his planetary model of the atom. At the beginning of his report he warned, "The stability of the atom proposed need not be considered at this stage." Again at the end he reiterated, "... the stability will obviously depend upon the minute structure of the atom and the motion of the constituent charged particles." Henceforth the Saturnian or nuclear model of the atom became known as the Rutherford model.

Although a great experimenter, Rutherford's presence in the lab could be a disturbance. The detection equipment was quite sensitive to vibrations and there was a large sign, "Talk softly please." In spite of this, Rutherford's booming voice could be heard vibrating through the walls.

Later in 1911, was the first Solvay conference that brought together the top physicists in the world. Rutherford was present. He was surprised that the continental physicists were not at all concerned with the internal structure of the atom. As he later wrote to his colleague, William Henry Bragg (1862 – 1942), "I was rather struck in Brussels by the fact that continental people do not worry their heads about the real cause of things."

Rutherford's discovery of the nuclear atom had an amusing consequence many years later in 1929. On a winter day of that year, at the Cavendish Laboratory, George Gamow (1904 – 1968), a physicist of Russian origin, was told to report to Rutherford who was looking for him. When Gamow reached Rutherford's office the latter rushed at him, thrust a piece of yellow stationery under his nose and demanded, "What does this mean?" The yellow piece of paper was a letter.

10 October, 1929
Rostow na Donu
U.S.S.R.
Dear Professor Rutherford,
We students of our university physics club
elect you honorary president because you
proved that atoms have balls.
Secretary Kondrashenkov.

Gamow could barely suppress his desire to laugh. Eventually he was able to mollify Rutherford by explaining to him that this was all simply a mistake resulting from a too-literal translation of the Russian words for atomic nucleus, namely *atomnoie yadro*. The word *yadro* can mean "kernel" or "ball" as in tennis ball.

Rutherford's experiments that showed that the number of electrons is small, was more difficult, but the implications were far-reaching. Neither the nuclear nor the plum-pudding atom was stable. In fact, the hydrogen atom was reduced to having only one electron, which, according to theory, would radiate all its energy away in a tiny fraction of a second. Thus, contrary to experience, almost no hydrogen should exist in the universe. And yet, hydrogen is the most abundant element. Again physicists were left with a terrible paradox and without any understanding of the stability of atoms.

Rutherford was the ideal mentor for a young physicist as the following statement illustrates. "I know nothing more deadening to original ideas than keeping a man's nose firmly fixed to the grindstone. Even directors need a change, and young men should have opportunities of meeting other young men working in other parts of the country. Ideas are more likely to come from such meetings with colleagues than by holding men down to some work in which there might be no progress at all. No laboratory today is self-sufficient."

"Experiment, directed by the disciplined imagination either of an individual or, better, of a group of individuals of a varied outlook, is able to achieve results which far transcend the imagination alone of the greatest philosopher."

Also even though Rutherford was the consummate experimentalist and Bohr a theorist, they became the best of friends. This in spite of the fact that unlike Rutherford, Bohr believed that, "When a person has mastered a subject thoroughly, he will then write so that hardly anyone can understand him." When Rutherford was asked about this remarkable friendship he replied, "Bohr is different; he's a football [2] player." Perhaps he thought that a soccer player would communicate clearly to an English barmaid.

About the time that Rutherford had established the Saturnian model, Bohr arrived in Manchester only to be confronted with these paradoxical results. He assaulted these obstacles squarely and had the temerity to suggest that the stability of atoms had to be simply accepted as an experimental fact to be used in the construction of a correct theory. So, just like Einstein who had turned the unexpected null result of the Michelson-Morley experiment into the expected, Niels Bohr turned the stability of atoms into a fundamental postulate. The stability of atoms was no longer to be viewed as due to their mechanical structure, but due to something else not yet understood.

Bohr drew up most of these results in notes (now referred to as the "Manchester Memorandum") during the summer of 1912. That Bohr considered these results important is revealed by a letter from 1912 to his brother in which he states, "It could be that I've perhaps found out a little bit about the structure of atoms. You must not tell anyone anything about it." As an interesting aside, the memorandum contained a serious error, which was essential for the solution of the problem of how to arrive at the periodic table. Without this error, Bohr's suggestion could not have hoped to compete with Thomson's model. Perhaps it was wishful thinking due to Bohr's anxiety to solve the problem. At any rate, the error went unnoticed.

On August 1, 1912, shortly after his return from his first period of work with Rutherford, Bohr married Margrethe Nørlund to whom he had been

---

[2]In Britain "football" is what in North America is called "soccer".

engaged in 1911. They spent a two-week honeymoon in Cambridge and then Niels returned to work with Rutherford.

The radically new ideas that Bohr proposed were that the planetary electrons could not revolve in an orbit with just any radius, but were confined to orbits with radii such that the energy of the orbit was given by Planck's constant times the frequency with which the electron revolved in its orbit. Secondly while in such an orbit, the electron was absolutely stable and did not radiate. Finally, an electron radiated only when it changed orbits, dropping from a higher to a lower orbit. The energy radiated is related to the frequency of the radiation by Planck's law. So in one fell swoop, the stability problem was solved and the discrete spectral lines were explained. The problem of the periodic table, however, was not solved until several years later and involved an entirely new concept, the Pauli Exclusion Principle.

After Bohr came up with his model, his school friend Hans Marius Hansen asked Bohr what his model had to say about spectra. When Bohr confessed that he had nothing to say about them, his friend suggested that he look at the Balmer formula. Years later Bohr stated, "As soon as I saw the Balmer formula, the whole thing was immediately clear to me." Most textbooks of physics today state that Bohr developed his model to explain the Balmer formula and the hydrogen atom spectrum.

Bohr earned the 1922 Nobel Prize in Physics "for his services in the investigation of the structure of atoms and of the radiation emanating from them".

In 1946, in his acceptance speech for the Nobel Prize, awarded to him for his Exclusion Principle, Wolfgang Pauli (1900 – 1958) referred to Bohr. "A new phase of my scientific life began when I met Niels Bohr personally for the first time. This was in 1922, when he gave a series of guest lectures at Göttingen, in which he reported on his theoretical investigations on the Periodic System of Elements. ... The question as to why all electrons in an atom in its ground state were not bound in the innermost shell had already been emphasized by Bohr as a fundamental problem. ... It made a strong impression on me that Bohr at that time and in later discussions was looking for a general explanation which should hold for the closing of every electron shell."

Using his assumptions, Bohr was able to obtain the Balmer formula for hydrogen. In fact, he obtained formulae for all the hydrogen series, many more of which had by now been observed. Thus, even though Bohr had not explained the stability of the Rutherford atom (in fact he just ignored this question) his success in explaining the hydrogen spectrum forced physicists, to pay attention to his work.

The first major public presentation of Bohr's model for the hydrogen atom occurred at the *Birmingham Meeting*, a yearly meeting of the British Association for the Advancement of Science. Lord Rayleigh was present, but refused to comment on the talk. As a grey eminence he was asked what he thought of the young Dane's ideas. He justified his silence with, "In my younger years I was convinced that a man over sixty should not take a part in a debate on new problems, and even though I no longer hold this view so firmly, I still hold it firmly enough not to want to take part in this discussion."

Somewhat earlier, when he first saw Bohr's paper, the seventy-one year old Lord Rayleigh had declared, "I have looked at it, but I saw it was no use to me. I do not say that discoveries may not be made in that sort of way. I think, very likely they may be. But it does not suit me."

Also in 1914 James Franck (1882 – 1964) and Gustav Hertz (1887 – 1975) at the Kaiser Wilhelm Institute in Berlin reported some unusual results from the scattering of electrons from mercury atoms. Gustav Hertz was the great-nephew of that Heinrich Hertz who had discovered the photoelectric effect. Bohr had actually suggested such an experiment in his paper, but Franck and Hertz had not seen his paper. They were simply looking for the "graininess" in atoms. In the 1960s James Franck recalled the situation as follows.

"It might interest you that when we made the experiments that we did not know Bohr's theory. We had neither read nor heard about it. We had not read it because we were negligent to read the literature well enough— and you know how that happens. On the other hand, one would think that other people would have told us about it. For instance we had a colloquium at that time in Berlin at which all the important papers were discussed. Nobody discussed Bohr's paper. Why not? The reason is that fifty years ago one was so convinced that nobody would, with the state of knowledge we had at that time, understand spectral line emission, so that if somebody published a paper about it, one assumed, 'probably it is not right.' So we did not know it."

After the Bohr model was published it was obvious that the Rydberg or Balmer series had provided an essential clue. When asked why these obvious regularities had not been used by anyone before him, Bohr replied, "They were looked upon in the same way as the lovely patterns in the wings of butterflies. Their beauty can be admired, but they are not supposed to reveal any fundamental biological laws."

Rutherford on receiving Bohr's manuscript wrote to him, "There appears to me one grave difficulty in your hypothesis, which I have no doubt you fully realize, namely, how does an electron decide what frequency it is going to vibrate at when it passes from one stationary state to another? It seems to me that you would have to assume that the electron knows beforehand where it is going to stop." In another letter to Bohr, Rutherford wrote, "No one can expect to clear up the whole of physics in a week, and one ought to feel grateful that the enterprise looks like going on forever."

At the University of Göttingen, where the first real quantum mechanics would be developed more than a decade later, Bohr's model was also not accepted. Here, the famous mathematician Richard Courant (1888 – 1972) was a friend of Niels Bohr's brother, Harald who was already quite famous. Thus, when Harald informed Courant that his brother Niels was really the more clever one, Courant was inclined to believe him in spite of the strange assumptions involved in the Bohr model of the atom. Most of the experimentalists at Göttingen, especially the spectroscopist Carl Runge (1856 – 1927), did not accept the Bohr model. Runge reacted quite strongly. "Now the literature on spectroscopy will be permanently contaminated with terrible things. ... This fellow is definitely mad." He went on to say, "Niels, it is true, has made a nice enough impression. But he obviously has done a strange if not crazy stunt with that paper, and if it were not for the wonderful brother Harald, one might just as well dismiss the whole so-called discovery." Later Courant wrote to Niels Bohr about this episode and to congratulate him on winning the Nobel Prize. "Thanks to a forewarning through Harald, who had so often told such amazing things about his brother, I was then at once prepared to believe that you must be right; but when I reported these things at Göttingen, they laughed at me for taking such fantastic ideas seriously; thus, as it were, I became a martyr to Bohr's model."

The reaction by Sir James Jeans (1877 – 1946) regarding the Bohr Theory was, "The only justification at present put forward for these assumptions is the very weighty one of success."

J. J. Thomson rejected Bohr's work as, "Hiding one's ignorance by invoking a new principle." This invoking of new "demons" to explain the unexplained seems to have reached a high point in the second half of the twentieth century, but was not yet commonplace in 1913. Anyway, to Thomson, Bohr's action was not physics but more akin to religion, in fact, a "cowardly substitute for knowledge of structure of the atom". Thomson tried hard to reinterpret Bohr's work within the framework of his own. Finally after World War One, Thomson realized that he was outside the mainstream of physics and resigned his chair at the Cavendish in favor of Rutherford. Today, such generous gentlemanly behavior seems to be extinct.

Einstein's reaction to Bohr's derivation of the hydrogen atom was more positive. What convinced him was a report from Rutherford. Certain spectral lines had been thought to belong to hydrogen but, according to Bohr's theory, belonged to helium. Experiments had confirmed that they belong to helium. Rutherford recounted in a letter to Bohr, "When I told him about the Fowler spectrum the big eyes of Einstein looked still bigger and he told me, 'Then it is one of the greatest discoveries.' I felt very happy hearing Einstein say so." At a later date Einstein again confirmed this opinion. "That this insecure and contradictory foundation was sufficient to enable a man of Bohr's unique instinct and perceptiveness to discover the major laws of the spectral lines and of the electron shells of the atom as well as their significance for chemistry appeared to me like a miracle and appears as a miracle even today. This is the highest form of musicality in the sphere of thought." Later in a letter to Ehrenfest Einstein wrote, "Bohr was here, and I am just as keen on him as you are. He is a very sensitive lad and goes about this world as if hypnotized."

"Rubbish! The electron in its orbit must emit radiation!" exclaimed Max von Laue. Einstein who happened to be attending the same meeting, however, defended Bohr's ideas, "No, this is remarkable! There is something behind it."

Even the future great, Otto Stern (1888 – 1969) who was then only twenty-five, later confessed to one of his students that he had vowed to drop physics if all this nonsense proved to be true. He laughingly admitted that when he and Max von Laue had taken their oath to drop physics they had tried to spoof a historical occasion. In 1307, on a hill called the Ruetliberg, the representatives of the Swiss cantons had sworn an oath to defend with all their might their independence from Austria. This was called the Ruelischwur (Ruetli oath). In 1914, on a hill called the Uetliberg, Max von Laue and Otto Stern also swore an oath to give up physics if and when these weird ideas of Niels Bohr regarding the hydrogen atom should prove correct. This they called the Uetlischwur (Uetli oath). Unlike the Swiss, they broke their vow.

All this shows is that new ideas are difficult to accept. It is indeed like Percy Williams Bridgman (1882 – 1961) said, "There is no adequate defense, except stupidity, against the impact of a new idea." He also stated, "I like to say that there is no scientific method as such, but that the most vital feature of the scientist's procedure has been merely to do his utmost with his mind, no holds barred."

Bohr himself was aware of the conflict with well-established theory that his ideas entailed. In a lecture to the Danish Physical Society in 1913 he did not shy from displaying his awareness of this difficulty. "Before closing I only wish to say that I hope I have expressed myself sufficiently clearly so that you have appreciated the extent to which these considerations conflict with the classical theory of electrodynamics. On the other hand, I have tried to convey to you the impression that just by emphasizing this conflict, it may be possible in the course of time to discover a certain coherence in the new ideas."

Even six years after World War One no Germans, except Einstein, were invited to the *Solvay Conference*. Einstein declined under these circumstances and even referred to it as "that witches' Sabbath in Brussels". Bohr also refused to attend unless these petty political hatreds were put aside.

By now, Bohr was one of the most famous physicists in the world and many professors sent their best students to spend some time at the Bohr Institute. So, when Bohr's friend Ehrenfest introduced Hendrik Bugt Casimir (1909 – 2000) in a letter he wrote, " ... *er kann schon etwas, aber braucht noch Prügel.*" (" ... he already has some abilities, but still needs thrashing.")

Ehrenfest could also not handle his young postdoctoral fellow J. Robert Oppenheimer (1904 – 1967), who quickly produced an answer to any question asked. Ehrenfest felt that some of the answers were wrong, but he could not find a reply quickly enough. So, he sent Oppenheimer to work with Pauli and wrote to the latter, "I have here a remarkable and intelligent American, but I cannot handle him. He is too clever for me. Could you take him over and thrash (*prügel*) him morally and intellectually into shape?"

In 1929 Ehrenfest took Casimir, his young student to a meeting at Bohr's Institute. Somewhere during the train ride there he told him, "Now you are going to get to know Niels Bohr and that is the most important event in the life of a young physicist." Casimir intended to stay for only two or three weeks. He remained for several months.

Casimir's father knew very little about physics and was not at all convinced that this chap Bohr with whom his son wanted to study was famous.

He became, however, quickly convinced when his first letter to his son addressed simply to:

H.B.G. Casimir

c/o Niels Bohr

Denmark

reached his son promptly and without any alterations of the address. This illustrates how well-known Bohr was in his native Denmark.

Even more revealing perhaps is the following episode. One evening Niels Bohr, Mrs. Bohr, the same Casimir and George Gamow were returning from a farewell dinner by O. Klein on his election as professor in his native Sweden. It was late and the streets of Copenhagen were deserted. On the way they passed a bank building made of large cement blocks that presented an irresistible temptation to an expert alpinist like Casimir. In no time at all he scrambled up to the third floor and back down. Not to be outdone Bohr, inexperienced though he was, decided to repeat the feat. As he had reached the second floor, with three worried people watching, two policemen noticed the suspicious activity around the bank and approached the group with their hands on their holsters. However, as soon as one of them had a good look at the "criminal" on the wall he said, "Oh, that's only Professor Bohr!" and off they went to look for more dangerous criminals.

The following anecdotes help to bring out some of the characteristics of this great man. In 1918, Bohr wrote to his friend O.W. Richardson about the difficulties of making further progress. "I know that you understand how things happen, and ..., from the scientific point of view, passes periods of overhappiness and despair, of feeling vigorous and overworked, of starting papers and not getting them published, because all the time I am gradually changing my views about this terrible riddle which the quantum theory is."

That Bohr had a childish streak is evident from his actions. In 1916 during World War One, Bohr and Kramers were almost killed. They were walking along a beach when they discovered a mine and like two little boys starting throwing rocks at it.

A beautiful "good-luck" horseshoe was nailed over the door of Bohr's cottage at Tisvilde. A surprised visitor asked, "Surely a great scientist like you doesn't believe that a horseshoe over the entrance brings luck?" "Of course not," answered Bohr, and then with a grin, "But you know, they say it works even if you don't believe in it."

It was virtually impossible to see Bohr separated from his pipe. Also like most pipe smokers, Bohr seemed to smoke more matches than tobacco. To allow for this he used to buy the biggest boxes of matches available and carry them with him wherever he went. In fact, he would frequently light his pipe before stuffing it with tobacco.

Another of Bohr's forms of relaxation was weeding. One day he was again occupied in weeding and had his pipe firmly clenched between his teeth when, unnoticed by him, the bowl fell off. His brother Aage Bohr and a young assistant Abraham Pais were lounging in the grass nearby and were treated to the wonderfully surprised look on Bohr's face when he tried to light a pipe without a bowl.

Later in his life, Abraham Pais (1918 – 2000) joined the Institute for Advanced Studies in Princeton where Einstein worked until his death. Einstein's office was rather large, and as far as he was concerned, too opulent and so he used the adjoining assistant's office as his own. This left his office available for important visitors. Consequently, on his annual visits to the Institute, Bohr was given Einstein's empty office. This led to the following rather amusing situation.

When working, Bohr required someone to bounce his ideas off. He would then pace about repeating a key word or phrase until his ideas were clearly formulated and then finally start discussing them with his colleague. Pais, as a former member of Bohr's institute in Copenhagen was well acquainted with this procedure and was usually called upon to work with Bohr when the latter visited Princeton. During one such session, in which Bohr was obviously preoccupied by Einstein's old objections to the interpretation of quantum mechanics, Bohr was pacing up and down muttering, "Einstein, Einstein, ..." Pais sat back waiting for the thought to jell and Bohr staring out the window continued to mutter, "Einstein, Einstein, ..." At this point, the door quietly opened and Einstein, whose doctor had forbidden him to smoke, sneaked in with a finger on his lips to Pais. Einstein was following his doctor's strictures literally: he had been forbidden to buy tobacco, but not to steal it. So here he was to get some of Bohr's. Just as Einstein reached Bohr's tobacco pouch on the desk the thought formed in Bohr's mind and with the shout, "Einstein" he turned around pointing his finger at the hapless Einstein caught with his fingers in the tobacco pouch. According to Pais, "For once, Bohr was truly speechless."

One of Bohr's great strengths was his ability to make intuitive judgements about complex situations, but sometimes he carried this too far. At a colloquium on nuclear isomers, in Princeton, Pais noticed that Bohr who was sitting beside him was getting more and more restless and kept whispering to him that it was all wrong. Finally, unable to contain himself any longer, he rose to make an objection, but stopped himself when he was only half up. He sat back down and turning to Pais with bewilderment asked, "What is an isomer?"[3]

It must have been quite an experience to drive with Bohr. More than one passenger has reported that during hot weather Bohr thought nothing of letting go of the steering wheel to remove his jacket. It was only through the intervention of the ever-alert Mrs. Bohr that the car was kept on the road during these occasions.

Bohr was a very kind and humane critic and would therefore generally try to ease the sting of negative criticism by some means. On one occasion he began by praising his opponent. His good friend Ehrenfest was also present and stopped him with the now famous phrase, *"Herr Bohr, fangen Sie gleich mit dem Aber an."* ("Herr Bohr, why don't you start right away with the but.")

Old style westerns, in which the good guys and bad guys were clearly distinguishable were Bohr's favorite movies. Frequently, if he got too tired working, he would ask who wanted to go to the movies and then drag some of his students along. Their purpose was to explain the more complicated aspects of the various plots to him. In these movies, the heroes naturally won all shoot-outs and Bohr developed a theory about shoot-outs to explain these successes. According to Bohr, the reason for the hero's success was purely psychological. The villain had to make a conscious effort to pull the gun. This thinking about what he had to do slowed the reflexes. The hero on the other hand simply had to respond to the villain's draw and was thus not slowed down. To test this theory Bohr's students bought a pair of holsters and toy guns so that Bohr could wear one set. For the next week his students tried to "shoot" him on encounters in the hall. Bohr's theory was verified experimentally since he invariably won. Later he derived a corollary. If the result of his theory were to become general knowledge and two gunfighters wanted to kill each other, neither one would be able to draw first and consequently the best they could do would be to talk.

After a particular western, Bohr exclaimed that he did not like the film because it was far too improbable. When asked to explain he said, "I believe that a beautiful girl might be walking along a dangerous trail in the mountains. I believe it possible that she might lose her footing and tumble over a cliff. I can also believe that she might grab hold of a small pine that just happened to be growing there. I can even believe that just at that moment a hero might come riding along and rescue her in the nick of time.

---

[3]In nuclear physics, an isomer is any of two or more nuclei having the same atomic number Z and mass number A, but different half-lives.

But that a camera should happen to also be there at that very moment to record all this I find just too improbable to believe."

Bohr analyzed the following simplest case in considering investments in the stock market. Consider people who only buy and sell reliable stocks so that fraud is not a consideration. Furthermore consider only the short-term effects of buying and selling in a fluctuating market. In such a situation, according to Bohr, a person who buys and sells completely at random is equally likely to win or lose, if we neglect things like taxes and broker's fees. However, those individuals connected with a company who have advance information about the company's success are able to make money on the stock market by using this information. Clearly for them to win, someone must lose. The only losers available are therefore those people who do not buy and sell at random and who have only incomplete information. They get their information from business reports etc. Thus, when they find a company that looks good and whose stock appears undervalued, they buy. This is exactly the wrong thing to do since those people with inside information must be selling to depress the price. Conclusion: If you have only limited knowledge then on the average you are more likely to lose than win on the stock market.

At the Carlsberg mansion, one evening, Bohr attempted to explain to the philosopher Harald Høffding the double slit experiment. Someone listening made the remark, "But the electron must be somewhere on its way from the source to the observation screen." To this Bohr replied, "What is in this case the meaning of the word 'to be'?" The philosopher Jørgen Jørgensen, who was also present, protested, "One cannot, damn it, reduce the whole of philosophy to a screen with two holes."

Bohr also had this to say about the difference between a scientist and a non-scientist. "We are no wiser and no less biased than other people. But as a physicist, or a biologist, you are certain to have gone through the experience of making a confident assertion, and then being proved wrong. A philosopher or a sociologist might never have had this wholesome lesson."

When Bohr visited Japan and gave a talk at Kyoto, there was a physicist named Kotani in the audience. Kotani was rather short and very slim so that he appeared rather youthful. During the question period following the talk, Kotani asked a question. Later Bohr was heard to comment, "That boy is extremely bright."

One of Bohr's favorite quotations was the last two lines from Schiller's *Sprüche des Konfuzius.* The full three lines read

Nur Beharrung führt zum Ziel,
Nur die Fülle führt zur Klarheit,
Und im Abgrund wohnt die Wahrheit.

Translation:

Only perseverance leads to the goal,
Only from fullness can clarity flow,
And truth resides in the abyss below.

Bohr also encouraged others by stating, "One must never be satisfied doing what one can; rather, one must always do what one really cannot." This should be contrasted with Wigner's "fatherly" advice to Walter M. Elsasser (1904 – 1991) when the latter's imagination had taken him a little too far. "One should tackle a problem only when its solution seems trivially easy. It will then turn out to be at the limits of the manageable. When it appears more difficult, trying to solve it is usually a hopeless thing."

In 1922, when Bohr was already very famous and had his own institute, but before Pauli had found the Pauli Exclusion Principle, necessary for explaining the periodic table, Bohr found a method to construct the periodic table. In this table he predicted that the element with atomic number 72 (Hafnium, named after the Latin name for Copenhagen, where it was identified in Bohr's laboratory by Coster and Hevesey) was not a rare earth, but rather a metal like zirconium.

After the discovery of Hafnium at the Bohr Institute, an open house was held to celebrate and arouse public interest. According to Pauli, Bohr was rushing around putting finishing touches to everything when he spotted Pauli standing alone in a corner. Bohr stopped and looked at Pauli for a while before commenting, "Pauli, you are more suitable to be exhibited than to exhibit."

Hafnium was found, using X-ray spectroscopy (Moseley's method), in zirconium ores and had chemical properties similar to zirconium. Henry Gwyn-Jeffreys Moseley (1887 – 1915) had published a paper, in late 1913 where he showed that the atomic number Z (the number of electric charges in an atom) could be easily measured by looking at certain specific lines in the X-ray spectra of the element. The frequency of these lines was simply related to an integer. He showed that the ordering of the elements in the periodic table was much more consistent if one used this atomic number rather than the mass of the atoms as Mendeléev had done. The difference was resolved much later after isotopes (atoms with the same atomic number but different mass) were discovered.

Harry, as Henry Moseley's father called him, was born into a scientifically active family. His father was a naturalist and professor at Oxford. His grandfather was a professor of divinity and his great-grandfather an unlicensed physician who had translated the bible into Chinese. It was inevitable that Harry went first to Eton and then to Oxford. At Eton the emphasis was on sports and leadership so that future leaders of the British Empire should emerge from its halls. This is where a deep sense of duty to one's country was instilled in the young man.

Moseley graduated from Trinity College, Oxford in 1910 and went as a lecturer to the University of Manchester. There he started his research career under Rutherford. He continued work along the lines of Max von Laue and W.H. Bragg, and W.L. Bragg on X-ray spectra of the elements. Sir William Henry Bragg and Sir William Lawrence Bragg shared the 1915 Nobel prize in physics "for their services in the analysis of crystal structure by means of X-rays".

Moseley in his laboratory.

Moseley's results proved to be a further beautiful verification of Bohr's theory when in 1916 Walther Kossel (1888 – 1956) gave an interpretation of these X-rays according to Bohr's theory. Kossel correctly assumed that these lines occurred when an inner electron had been ejected from an atom and an electron from further out jumped into this lower orbit. In this way it was possible to determine where exactly in the periodic table a given element belonged. Moseley was able to show where there were gaps in the periodic table. This also settled one of the questions that had stumped chemists for decades, "How many rare earth elements are there?"

The French chemist, Georges Urbain (1872 – 1938) was amazed when he brought some samples of rare earth elements, on which he had labored for years, and Moseley analyzed them in a matter of hours. Urbain wrote to Rutherford, "I was most surprised to find a very young man capable of doing such remarkable work. ... Moseley's law, for the end as well as for the beginning of the group of rare earths, has established in a few days the conclusions of my efforts of 20 years of patient work. However, it is not only that which makes me admire Moseley's work. His law replaced the somewhat imaginative classification of Mendeléev with one, which was scientifically precise. It brought something definite into a period of the hesitant research on the elements. It ended one of the finest chapters in the history of science."

Later, Moseley realized that his law is a direct consequence of the Bohr formula. In 1914 he stated, "I should be glad to do something towards knocking on the head the very prevalent view that Bohr's work is all juggling with numbers until they can be got to fit. I myself feel convinced that what I have called the 'h' hypothesis is true, that is to say one will be able to build atoms out of e, m, and h and nothing else besides. Of the three variations of this hypothesis now going, Bohr's has far and away the most to recommend it, but very likely his special mechanism of angular momentum and so forth will be superceded."

In 1914, when World War One broke out, Moseley volunteered for king and country and was commissioned in the Royal Engineers as a signals officer. His unit was sent to the Dardanelles to participate in the ill-fated Gallipoli campaign. There, two guides that later disappeared, misled his brigade. The soldiers were left in a position ahead of their own lines. In the morning, when they woke up, the men realized where they were. The Turks had started their attack. Moseley, only twenty-seven years old, was shot through the head, and the world lost a great scientist in a senseless campaign.

As Robert Millikan lamented, "In a research which is destined to rank as one of the dozen most brilliant in the history of science a young man twenty-eight [sic] years old threw open the windows through which we can glimpse the sub-atomic world with a definiteness and certainty never dreamt of before. Had the European War had no other result than the snuffing out of this young life, that alone would make it one of the most hideous and irreparable crimes in history. ..."

Chapter 8

Action in Physics: The Old Quantum Theory

"The quantum theory is much like some victories, for a month or two you are laughing and then you cry for long years." Hendrick Kramers

Some of Bohr's results had actually been foreshadowed by Arnold Johannes Wilhelm Sommerfeld (1868 – 1951). Here was a man totally committed to science. Just prior to the appearance of Einstein's paper on Special Relativity, he had finished some outstanding work on the electrodynamics of electrons in motion. He had even considered electrons moving at speeds greater than light. When Einstein's paper on special relativity appeared and showed that such motion was not possible for electrons in empty space, Sommerfeld abandoned his work and turned to relativity.

He did however state, "The term 'Theory of Relativity' is an unfortunate choice. Its essence is not the relativity of space and time, but rather the independence of the laws of nature from the viewpoint of the observer. The bad name has led the public to believe that the theory involves the relativity of ethical conceptions somehow like Nietzsche's *Beyond Good and Evil*." Both Felix Klein (1849 – 1925) and Sommerfeld suggested the name "Theory of Invariants", but this never caught on.

An interesting later development is that Sommerfeld's work for electrons moving at speeds greater than the speed of light was rediscovered for light propagating in materials with a sufficiently high index of refraction where electrons can indeed move at speeds above the speed of light in that material.

In 1934, Pavel Alexeyevic Cherenkov (1904 – 1990) discovered this kind of radiation, now named after him. The radiation shows the conical shock wave discovered by Sommerfeld in 1904. Cherenkov shared the 1958 Nobel Prize in Physics with Il'ja Mikhailovich Frank (1908 – 1990) and Igor Yevgenyevich Tamm (1895 – 1971) "for the discovery and the interpretation of the Cherenkov effect".

Sommerfeld's father was a physician in Königsberg (present-day Russia's Kaliningrad). In his biographical note Arnold Sommerfeld wrote, "My father ... was a passionate collector of natural objects (amber, shells, minerals, beetles, etc.) and a great friend of the natural sciences. ... To my energetic and intellectually vigorous mother I owe an infinite debt." He attended the Altstädtische Gymnasium in Königsberg. Two future colleagues, Hermann Minkowski and Willi Wien, were just ahead of him in this high school. Arnold graduated with excellent grades in 1886 and entered the University of Königsberg where some of the top mathematicians (two of whom we meet later), David Hilbert (1862 – 1943), Adolf Hurwitz (1859 – 1919), and Ferdinand von Lindemann (1852 – 1939) taught. The lectures by Hilbert convinced Sommerfeld to concentrate on mathematics.

It was fortunate that Sommerfeld had such inspiring teachers because he could easily have neglected his studies more than he did after he joined a *Bruderschaft*. As a member of the fraternity, he was expected to take part in drinking bouts and dueling. For the rest of his life Sommerfeld bore, as a mark of his dueling prowess and memento of his student days, a long scar on his forehead. Contrary to the fashion of the time, Sommerfeld finished all of his studies at Königsberg and received his doctorate under von Lindemann in 1891. He remained there for another year to receive a teaching diploma. In 1892 he fulfilled his military obligations by joining a reserve regiment in Königsberg. Unlike other academics, Sommerfeld seems to have enjoyed his military activities and volunteered every summer for the next eight years to take part in military exercises. Following his year of military service he moved to Göttingen, then "the seat of mathematical high culture". Here he met Felix Christian Klein (1849 – 1925) whom he came to consider his real teacher not only in mathematics, but also in all of mathematical physics. He became Klein's assistant and received his *Habilitation* (the right to teach at a university) under him. He also started with Klein a 13-year study of gyroscopes, culminating in the monumental four-volume *Über die Theorie des Kreisels* (On The Theory of the Top). The first two volumes deal with the mathematical theory and the last two volumes handle applications to geophysics, astronomy, and technology. This work is of importance even today to anyone, like NASA, using gyroscopes for navigation.

After Göttingen, Sommerfeld accepted positions as professor at the Clausthal Mining Academy followed by the Technical School in Aachen.

Although not the most exciting of positions, at Clausthal Sommerfeld was no longer a Privatdozent (an unpaid lecturer whose income came strictly from what the students paid him) and had sufficient income that he could marry Johanna Höpfner, the daughter of the University's curator. Finally in 1906 with support from Klein he obtained the chair of theoretical physics at Munich where he remained.

Sommerfeld was also one of the world's greatest teachers of mathematical physics. One of his favorite techniques was as follows. During seminars he would listen quietly to the talk presented by a student until he felt that the student was skipping some important detail. Then he would say, "I don't understand" and force the student to give a clearer explanation. If the student could not, then Sommerfeld himself would explain. He also once told a bright student, "Well, I can't solve this problem, now you try it." When asked how he could lecture on a subject he did not understand he responded, "If I knew something about it I would not lecture on it."

Arnold Sommerfeld and Niels Bohr in Lund, Sweden.

This is how Max Born explained Sommerfeld's teaching. "Theoretical physics is a subject which attracts youngsters with a philosophical mind who speculate about the highest principles without sufficient foundations. It was just this type of beginner that he knew how to handle, leading them step by step to a realization of their lack of actual knowledge and providing them with the skill necessary for fertile research. ...He had the rare ability to have time to spare for his pupils, in spite of his duties and scientific work. ...In this friendly and informal way of teaching a great part was played by invitations to join a skiing party on the 'Sudelfeld' two hours by rail from Munich. There he and his mechanic ...were joint owners of a ski hut. In the evenings, when the simple meal was cooked, the dishes were washed, the weather and snow properly discussed, the talk invariably turned to mathematical physics, and this was the occasion for the receptive students to learn the master's inner thoughts."

According to one of Sommerfeld's students, Rudolf Peierls, his teacher was famous for making silly mistakes while lecturing, but like all good theorists he could always find an argument to get the right answer. As an example, Peierls told the following story. Sommerfeld was lecturing on a theory of electrons in metals. At the end of a calculation he realized that he was out by a factor of two. To get the right answer he then stated, "Half the electrons go one way and half the electrons go the other way, so we have to divide by two."

It was also quite common for Sommerfeld to visit the Hofgarten Café both before and after a colloquium to discuss problems with a coworker. Consequently the marble surface of the table was frequently covered with mathematical formulas. On one occasion he was unable to work out an integral, before returning to his office, and left it on the table. The next day when he returned, the solution was written below his formula. A mathematician colleague had taken his time at the cafe and worked out the integral.

The quality of Sommerfeld's teaching is evident from the fact that almost all of his students received the Nobel Prize. He himself, however, did not. It seems that after World War One, resentment against Germany and German physicists was too great.

Incidentally, after World War One it was necessary to get a separate visa for every country in Europe. One physicist, Robert Brode, who was studying in Germany, devised an ingenious way to avoid the long lineups at the various consulates. He and his colleagues had visiting cards printed, which stated under their name "Member, International Adiabatic Commission". They would present these cards to a clerk at the consulate and ask him to transmit it to the consul. Without exception they were immediately ushered into the presence of the consul and issued a visa, without charge. No one ever asked what the function of that commission was.

Like Planck, Sommerfeld was a mature, well-established and highly respected professor who had made numerous important contributions before bringing atomic physics to one of its high points. A somewhat stuffy, stereotypical German professor, he was a very warm and courageous human being. During the dark days of the Nazi regime he stood up for Einstein and other Jewish scientists. As early as 1911 he declined an invitation to give a talk on Special Relativity since he considered it a "secured possession" of physics and no longer a research problem of current interest.

When Hitler's policies caused a mass exodus of great young physicists from Europe to America, the aging Sommerfeld lamented the fact in a moving letter with the words, *"Sic transit gloria mundi veneris ..."* (Thus, passes the glory of the Old World to the greater glory of the New World.)

In a letter to Einstein written in 1934, Sommerfeld had this to say. "Moreover I can assure you that the misuse of the word 'national' by our rulers has thoroughly broken me of the habit of national feeling that was so pronounced in my case. I would now be willing to see Germany disappear as a power and merge into a pacified Europe."

He further wrote, "It may interest you that in the winter in my general lectures on electrodynamics I allowed them to ring out and crowned them with a summary of the special theory of relativity. The students were enraptured, because I also was once more enraptured by the beauty and unity of the system; not one was opposed."

Also in 1934 Sommerfeld gave a lecture in Holland and received a very handsome fee. With almost childish glee he told Emilio Segré (1905 – 1989), "I'm sending this immediately to Rutherford for the benefit of displaced scholars. I cannot do so from Germany, but this is an opportunity I will not miss."

There is an amusing story, about his stuffiness, from the time before Hitler. An American student who had arrived at Sommerfeld's institute could not understand the Herr Professor's coolness toward him. Finally, a fellow student explained to him that he should not address Sommerfeld as, "Herr Professor", but rather as, *"Herr Geheimrat"*, literally, "Privy-councillor", a meaningless title that was awarded by the state to distinguished people. At the next occasion the student did just that and was rewarded by a warm smile from Sommerfeld who proclaimed, "Ach, your German is making wonderful progress."

After Bohr published his model of the hydrogen atom, Sommerfeld was lured back to an idea he had published a few years earlier in the *Proceedings of the German Physical Society*. There he had pointed out that since Planck's constant had the dimensions of action, the classical action of a system should be just a multiple of Planck's constant. As it turned out, this hypothesis not only implied Bohr's condition on the orbits of atoms but, was an elegant and far-reaching generalization of Bohr's assumption. At the time that he first made the suggestion Sommerfeld was too occupied to follow it up. Bohr's work revived Sommerfeld's interest and he returned to the idea.

Sommerfeld now used his own, more elegant formulation, and showed that in addition to Bohr's circular orbits it was also possible to have elliptic orbits. He went on to compute the effect of elliptic orbits as well as corrections due to relativistic effects. All of these computations gave small corrections to Bohr's original formula and were beautifully verified by experiment. This paper was published in 1916 at the height of World War One.

In spite of the war, Rutherford and Sommerfeld continued to correspond via Niels Bohr who managed somehow to get the letters through the sealed borders. As Rutherford said, "If we write at all, it will be worthwhile to see that physics is not left out in the cold." This is the same Rutherford that years later, on a matter of principle, refused an invitation to Max Born's home because Fritz Haber, the inventor of gas warfare, was going to be present.

Bohr-Sommerfeld quantization—as the use of Sommerfeld's action principle approach was called— led to the result that only certain directions were allowed for the magnetic moment of a particle. It was as if the little magnet associated with a particle could only point in very few directions. For an atom of angular momentum one-half, only two directions were allowed. For an atom of angular momentum one, three were allowed, and so it went. This so-called "space-quantization" was very controversial. Pieter Debye (1884 – 1966) told Walther Gerlach (1889 – 1979), "Surely you don't believe that this orientation of the atoms is something physical, it is a prescription for calculating—a lecture book for electrons."

Even Sommerfeld vacillated between the two opinions: reality and prescription. Bohr, however, believed in the physical reality of the three-fold orientation of the atom with angular momentum one, but thought that for stability reasons the direction perpendicular to the magnetic field had to be omitted.

Otto Stern (1888 – 1969) concluded that it was absolutely essential to test the matter experimentally. Although a theorist, he was full of experimental ideas and suggested how to do this. He and Max Born had already experimented with atomic beams. In the fall of 1920 Walther Gerlach arrived at the neighboring institute. A very good experimenter, he once stated, "The right professor isn't someone who knows everything better, but rather the one who is himself clear where and when there is something he does not know." Gerlach's arrival led to the famous Stern-Gerlach experiment, which proved that "space-quantization" was real.

Robert Andrews Millikan and Otto Stern.

The funding for this experiment had been almost wiped out by the inflation sweeping Germany after World War One. However, after Eddington's expedition to the island of Principe where the observation of the solar eclipse gave excellent verification of Einstein's prediction of light deflection, an Einstein craze swept the world. Everybody wanted to know what the prediction and the experiment were about.

Born used this fact to his advantage to finance the experiment. He announced a series of three lectures on Einstein's Theory in the biggest lecture hall of the University. The event was a colossal success. The hall was crowded and a large sum was collected. Born's business friends advised him that he had gone about this all wrong. He would, according to them, have done much better if he had sent out private invitations to lecture in the most expensive hotel in evening dress, with cocktails to be served, and then had asked for an assistance fund. That, however, was not Born's style.

Another success of Sommerfeld's formulation was an explanation of the Stark effect. Johannes Stark (1874 – 1957), had discovered a phenomenon similar to the Zeeman effect. In the Zeeman effect if an atom is placed in a strong magnetic field, the lines in the spectrum of the atom split. In the Stark effect when an atom is placed in a strong electric field, its lines also split. Bohr had been unable to explain this effect, but Paul Epstein and Karl Schwarzschild independently used Sommerfeld's refined formulation and again the results of their calculation agreed beautifully with the experiment.

Later during the Nazi period Johannes Stark, a staunch Nazi, became president of the Physical-Technical Reichinstitut in Berlin. Stark, whose abilities were limited, considered himself also a spectroscopist. As one of his first experiments he suggested that they look for the ground state of orthohelium. The spectroscopist, Hans Kopfermann (1895 – 1963) knew from quantum mechanics that this ground state did not exist. However, he also knew that Stark did not accept quantum mechanics, which he called "Jewish physics". Thus, Kopfermann looked off into space and said, "Herr President, I have a feeling that that state does not exist." Thus, when Stark and his colleagues failed to find this state, Kopfermann was considered by them to be a great scientist because he had "the right German feeling" for how to do physics.

Once Sommerfeld's quantization condition became general knowledge, physicists applied it to ever more complex atoms. They now had a theory and set to with a frenzy, trying to understand all of atomic structure. The results were beautifully summarized in Sommerfeld's four separate and continuously revised editions of *Atombau und Spektrallinien* published by him first in 1919 with the last issue in 1924. In the Preface to *Atombau und Spektrallinien* he rhapsodized, "Since the discovery of spectral analysis no one in the know can doubt that the problem of the atom will be solved if we have learned to understand the language of the spectra. ... What we hear from the language of the spectra these days is truly a music of the spheres of the atom, a consonance of integer relations, an increasing order and harmony under all manifold circumstances. ... All integer laws of the spectral lines and atomic theory flow in the final analysis from the quantum theory. This is the mysterious organ on which nature plays the spectral music and according to whose rhythm she organizes the construction of the atoms and the nuclei."

He also, with some irony, called Bohr's correspondence principle, "a magic wand" because one could do calculations with it although it had no rigorous mathematical formulation.

In 1917, Einstein published a paper pointing out the difficulties arising in Bohr-Sommerfeld quantization due to what today is called chaos. This paper was ignored for several decades. Even the famous are ignored if their ideas are premature.

Sommerfeld's books epitomize the high point of the Old Quantum Theory. It was a theory that was applicable to all periodic motions and when applied to these was highly successful. The fifth, planned edition, however, was never published because in the span of one short year 1925 – 1926 the Old Quantum Theory came to be replaced by what is the modern theory of Quantum Mechanics.

After the war, Sommerfeld, although 77 years old, helped to rebuild the intellectual climate in Germany. He continued to lecture until 1947 and worked hard to publish his six-volume lectures on theoretical physics. Fate intervened and he never finished the final volume. Sommerfeld was walking with his grandchildren and, being a little deaf, did not hear a warning. A truck struck him and two months later he died from the injuries.

Chapter 9

Particles are Waves are Particles

"Are not the rays of Light very small Bodies emitted from shining Substances?" Isaac Newton, Philosophical transactions (1672).

At the beginning of the twentieth century, physicists again learned to view optical theory in an earlier light. In the seventeenth century Newton asserted that light was corpuscular or particle-like in nature. A century and a half later the work of luminaries like Thomas Young (1773 – 1829) and Augustin Jean Fresnel (1788 – 1827) definitely showed that, contrary to Newton's view, light was wave-like and not corpuscular, or particle-like, in nature. By passing light through a slit they demonstrated that light was diffracted just like a sound wave or water wave. It was only the very short wavelengths of visible light that had made it appear to travel like a particle in a straight line without spreading. This viewpoint was about to be reversed.

In 1887, at the technological institute at Karlsruhe, Heinrich Rudolf Hertz, only thirty years old, a master of Homer and Greek tragedies, as well as economics and the history of mathematics and physics, was about to open the doors to wireless communication and radio. During the four short years between 1885 and 1889, he not only demonstrated that accelerating charges produce radiation that travels at the speed of light but that this radiation conforms in every respect to the radiation predicted by the theory of electromagnetism produced by James Clerk Maxwell twenty years earlier. After he produced the first electromagnetic (radio) waves, his students were impressed and asked him, "What next?" he simply replied, "It's of no use whatsoever. This is just an experiment that proves Maestro Maxwell was right—we just have these mysterious electromagnetic waves that we cannot see with the naked eye. But they are there."

Hertz was born in Hamburg where his father was a prominent lawyer and later senator. He passed his university entrance exams at the top of his class. After a year of military service (1876 – 1877) he went to the University of Munich for a year, but decided that instead of engineering he wanted to pursue an academic career in science. In 1878, he moved to the University of Berlin where he studied with Hermann von Helmholtz and received his Ph.D. magna cum laude in 1880 with a thesis on electromagnetic induction. As Helmholtz' favorite pupil, Hertz remained with him as his assistant for three more years.

In 1883, he became lecturer in theoretical physics at Kiel and only two years later was called to the chair of physics at Karlsruhe where he remained until 1889. Here he met Elizabeth Doll, the daughter of one of the professors. A year later they married. At Karlsruhe he set out to test which of the competing theories of electrodynamics: Maxwell's or Helmholtz's was correct. Although he leaned towards his teacher's theory, he showed unequivocally that Maxwell's theory was correct. Regarding Maxwell's equations he later stated, "One cannot escape the feeling that these mathematical formulae have an independent existence and an intelligence of their own, that they are wiser than we are, wiser even than their discoverers, that we get more out of them than was originally put into them." Even the great Lorentz was converted from Helmholtz's electromagnetic theory by Hertz's experiments, which he felt were "the greatest triumph that Maxwell's theory has achieved."

To produce radio waves, Hertz constructed the world's first transmitter consisting of a high-voltage coil and a spark-gap. The receiver was an antenna consisting of a wire with a tiny spark gap across which he applied a voltage slightly too low for a spark to occur. When radio waves from the transmitter (produced with the spark) struck the antenna, they induced a small electric current and produced a spark across the gap.

The importance of Hertz's work was recognized immediately. A few days after Hertz's death, Boltzmann wrote to Helmholtz, "One should emphasize the extraordinary import of Hertz's discoveries in relation to our whole concept of Nature, and the fact that beyond a doubt they have pointed out the only true direction that investigation can take for many years to come." Similarly, Sir Oliver Lodge credited Hertz with accomplishing what the great English physicists could not. "Not only had he established the validity of Maxwell's theorems, he had done so with a winning modesty."

These primitive sparks were the world's first artificially produced and detected radio waves. The careful Hertz, however, noted one extra fact. If light from the transmitter spark shone on the spark gap of the receiver, the receiver sparked with an even slightly larger gap. It was as if the air had been made more conducting due to the incident light. This simple observation of the so-called photoelectric effect eventually led to the conclusion that light was not a wave at all, but rather consisted of corpuscles or quanta of energy.

The point was that the air was made more conducting by the ejection of electrons from the metal wires. This caused a serious problem for the physics of the time because, if light was a wave, then its energy would have to be absorbed by the metal surface as a whole. The energy from the wave would be smeared out over the whole surface. This meant that for electrons to be ejected, enough light would have to hit the metal surface to eject *all* its surface electrons. With light sources then available this would have required centuries of waiting for his phenomenon to happen. It should never have been seen, yet there it was.

In 1905, Einstein's miracle year, in order to explain this photoelectric effect, he proposed that light should be particle-like. If such a particle, a photon, hits an electron near the surface of a metal, it can give up all its energy to this one electron like one billiard ball hitting another. So, a single photon with enough energy (given by Planck's constant times the frequency of the photon viewed as a wave) can eject an electron from the surface. According to this view the ejected electron would have kinetic energy equal to the difference between the photon energy and the energy that kept the electron tied to the metal. In 1909 he repeated, "It is undeniable that there is an extensive group of data concerning radiation which show that light has certain fundamental properties that can be understood much more readily from the standpoint of the Newtonian emission theory than from the standpoint of the wave theory. It is my opinion therefore that the next phase of the development of theoretical physics will bring us a theory of light that can be interpreted as a kind of fusion of the wave and emission theory."

The term "photon" is due to the physical chemist Gilbert Newton Lewis (1875 – 1946) who also suggested the name "jiffy" for the time it takes light to travel one centimeter.

Robert Millikan (1868 – 1953) verified Einstein's explanation of the photoelectric effect in the period 1912 to 1915. It was an exceedingly difficult experiment because the energy with which the electron is bound to the surface of the metal depends very sensitively on any surface contamination.

You need very clean surfaces to get a reproducible result. Millikan solved this by enclosing a chunk of sodium metal, which is very soft, inside a vacuum. He used a razor blade that he could manipulate from outside the vacuum to shave a thin slice off the surface after each measurement so that the photons always hit a fresh surface. Millikan succeeded not only in verifying Einstein's equation (that the kinetic energy of the emitted electrons is given by Planck's constant times the frequency of the light less the energy with which the electron is attached to the surface), but he also obtained the most accurate measurement of Planck's constant up to that time.

Millikan hailed from pioneering stock in Morrison, Illinois. After high school he briefly worked as a court reporter before enrolling in Oberlin College, Ohio where, due to a lack of competent teachers, he taught himself physics. His favorite subjects were Greek and mathematics. After a master's degree at Columbia University, he received a fellowship in physics and earned a Ph.D. in 1895. For a time, Millikan was the sole graduate student in physics at Columbia. At the advice of his professors, he spent the year 1895-96 at the Universities of Berlin and Göttingen.

He returned to the U.S. in 1896 to work as an assistant to Michelson at the University of Chicago. To do so, he had given up much more lucrative offers because Michelson had promised him he could spend half his time on research—a rare opportunity. In 1902, he married Greta Blanchard and in 1910 became professor of physics at Chicago. He remained there until 1921 except for an interlude in 1917 when he went to Washington to work on war research on "the detection of submarines and other essential problems." There he met the astronomer George Hale (1868 – 1938) who convinced him to move to the California Institute of Technology in Pasadena, California as director of the Norman Bridge Laboratory of Physics.

Later Millikan became involved in a rather unpleasant controversy with Arthur Holly Compton about the nature of the recently discovered cosmic rays. Eventually Millikan was forced to concede that Compton was right.

Ironically, although he verified Einstein's equation, Millikan did not accept Einstein's interpretation that this showed that photons were particle-like. In 1912 at a joint meeting of the American Association for the Advancement of Science and the American Physical Society he stated quite unambiguously that a corpuscular theory of light was for him "quite unthinkable", with the phenomena of diffraction and interference. In 1923, he was the second American (after Michelson) to receive the Nobel prize in physics "for his work on the elementary charge of electricity and on the photoelectric effect". During his Nobel address he stated, "This work resulted, contrary to my own expectation, in the first direct experimental proof ... of the Einstein equation and the first direct photo-electric determination of Planck's *h*." He refrained from stating that this demonstrated the corpuscular nature of light.

Finally, in 1950 at age 82 in his autobiography, Millikan's memory suffered a revision. He wrote that his experimental data "proved simply and irrefutably, I thought, that the emitted electron that escapes with the energy *hf* gets that energy by the direct transfer of *hf* units of energy from the light to the electron, and hence scarcely permits of any other interpretation than that which Einstein had originally suggested, namely that of the semi-corpuscular or photon theory of light itself."

The part of Millikan's Nobel citation referring to "for his work on the elementary charge of electricity" goes back to earlier work. In 1910, Millikan had been the first to accurately measure the electric charge on an electron and show that all charges were a multiple of this elementary charge. This was his famous oil drop experiment which generations of physics students have been forced to repeat. In this experiment a tiny oil drop with a few electron charges on it is kept from falling by the application of an electric field. By balancing the force due to gravity, the electric force and the resistance due to air, Millikan could show that all the drops carried a charge that was a simple multiple (1, 2, 3, 4, etc.) of some elementary charge. His results for this elementary charge were in good agreement with the much earlier values that Planck had obtained in 1900 for the charge on an electron from his formula for blackbody radiation.

In 1922, Arthur Holly Compton (1892 – 1962) also demonstrated, in a totally new way, that light definitely has a particle nature. In a series of brilliantly beautiful experiments he bounced X-rays off electrons and showed that the collisions were in every respect like those of point-like, colliding billiard balls. These experiments are called the "Compton effect". Compton shared the 1927 Nobel Prize in physics "for his discovery of the effect named after him" with Charles Thomson Rees Wilson (1869 – 1959) "for his method of making the paths of electrically charged particles visible by condensation of vapor".

Compton received his B.Sc. from the College of Wooster in Ohio where his father was Dean and Professor of Philosophy. Arthur Compton considered pursuing a religious career, but his older brother Karl (a physicist) got him interested in X-rays and his father advised him to go into science. "Your work in this field may become a more valuable Christian service than if you were to enter the ministry or become a missionary." Arthur followed

Karl to Princeton University where he finished a Ph.D. in 1916 studying the angular distribution of X-rays reflected from crystals. He continued to study the scattering of X-rays by crystals even after that. This led him to his famous result. He found that the wavelength of X-rays increased when scattered off free electrons. This meant that the X-rays had lost energy during the scattering. He then considered the X-rays as particles and used conservation of energy and momentum as for ordinary billiard balls and found that the change in energy of the photon depended on the angle at which it bounced off the electron. This calculation and his experimental results agreed beautifully.

In 1920, Compton presented his preliminary results to the Cambridge Physical Society. Rutherford introduced him as follows. "This is Dr. Compton who is with us from the United States to discuss his work on 'The Size of the Electron'. I hope you will listen to him attentively, but you don't have to believe him."

Conference on Nuclear Physics in Rome, 1931.
The four prominent people in front are, left to right: Robert Andrews Millikan, Arthur Holly Compton, Marie Curie and Guglielmo Marconi. Also visible at the back are: Max Theodor Felix von Laue and Paul Ehrenfest.

In 1934 when Compton spent a year as Visiting Professor at Oxford, Rutherford forwarded some letters that had been mistakenly sent care of him to Cambridge. At that time science was somewhat neglected at Oxford. Rutherford had attached a note to the letters. "Dear Compton, I hear that you are at Oxford. Whatever can you be doing there?"

As Rutherford's comments imply, Compton's findings were not immediately accepted despite hints of similar results due to Sir William Henry Bragg whose son, Sir William Lawrence Bragg, had this to say about Compton's results: "God runs electromagnetics on Monday, Wednesday, and Friday by the wave theory, and the devil runs it by quantum theory on Tuesday, Thursday, and Saturday." He also had this to say, "The important thing in science is not so much to obtain new facts as to discover new ways of thinking about them."

Compton's chief opponent was William Duane of Harvard University. Duane accepted photons in other X-ray phenomena, but refused to accept them in scattering experiments mainly because he could not repeat Compton's results in his own laboratory. At an American Physical Society meeting, Duane attacked Compton and in Compton's words, "I might have criticized his interpretation of his results on rather obvious grounds, but thought it would be wiser to let Duane himself find the answer." Later Compton visited Harvard, but the dispute remained unresolved.

Another example of the kind of resistance he encountered is illustrated by the following. At a meeting of the British Association for the Advancement of Science in Toronto, a lengthy debate ensued after Compton presented his results. Following the debate, the future Nobelaureate, Chandrasekhara Venkata Raman (1888 – 1970) walked up to him and stated, "Compton, you are a good debater, but the truth is not in you."

It seems that when Compton received the Nobel Prize in 1927 for his work with X-rays this acted as an extra stimulus for Raman. He thought that his newly discovered radiation (now called the Raman effect) might be a similar effect with visible light. Raman received the 1930 Nobel Prize "for his work on the scattering of light and for the discovery of the effect named after him".

Raman's interest in light scattering started on his voyage home to Calcutta from the *Congress of Universities of the British Empire* held at Oxford in the summer of 1921. While on board, in the Mediterranean, he tested Lord Rayleigh's notion that the blue of the sea was due to reflection of the blue of the sky and found this to be wrong. Using a simple arrangement of crossed Nicol prisms (the polaroid lenses of his time) he could eliminate the polarized reflected light of the sky and found that the blue of the water persisted as a much deeper blue than that of the sky. This started his experiments with very clean water after he got home.

Later Raman became one of the great stars of Indian physics. Thus, when the India's Education Minister, M. C. Chagla, offered the Government's financial support he responded, "Sir, I want this Institute to be an oasis in the desert, free from government interference and the application of its rules and regulations. That would destroy my Institute. Thank you for your offer."

In a similar vein he responded to Prime Minister J. Nehru when this statesman admonished India's scientists and asked them to emerge from their ivory towers. "The men who matter are those who sit in ivory towers. They are the salt of the earth and it is to them that humanity owes its existence and progress."

As a good Hindu, Raman was a teetotaler. At a banquet in Bordeaux where he was the guest of honor, Professor J Cabannes proposed a toast and everyone raised a glass of wine except Raman who raised a glass of water with the words, "Sir, I know what my effect on alcohol is, but I certainly don't want to try the effect of alcohol on me."

When Herbert Skinner (1897 – 1958) in August 1923 gave a talk on the Compton effect to the Kapitza Club in Cambridge no one believed the result. The upshot was that many of the prominent physicists signed a note in the minutes book stating, "Compton is wrong." The first four signatories were: Piotr Kapitza (1894 – 1984), Patrick Blackett (1897 – 1974), as well Douglas Rayner Hartree (1897 – 1958) and Skinner. Blackett was converted to Compton's view only later, after a discussion of Pauli's theory of radiative equilibrium which was also based on collisions between light and particles.

Thus, after Compton's results were accepted, light (in this case X-rays) had to be both a wave and a particle. The physicists' world had become schizophrenic indeed. Regarding the wave-particle duality, J. J. Thomson stated that conflict between the two is like, "... a struggle between a tiger and a shark, each is supreme in his own element, but helpless in that of the other."

Before a resolution to this dilemma could be found, the physicists' world had to become even a little more topsy-turvy. Objects such as electrons, which had definitely been established as particles, had to be turned into waves. This step was taken by Prince Louis-Victor Pierre Raymond de Broglie (1892 – 1987), younger brother of Duc Maurice de Broglie. They were scions of one of the older noble houses of France. Since the seventeenth century the family had produced generals, politicians and diplomats. On his brother's death Prince Louis-Victor inherited the title of Duc.

In 1911 the first Solvay Congress brought together many of the world's greatest minds to discuss the difficulties presented by the theory of radiation and quanta. Prince Louis-Victor de Broglie's older brother, Maurice (1875 – 1960), was one of the scientific secretaries at the meeting and the young prince helped his brother with the work of editing the proceedings for publication. The spirited discussion of Planck's quanta so impressed the young prince that as he later stated, "I decided to devote all my efforts to investigate the real nature of the mysterious quanta that Planck had introduced into theoretical physics ten years earlier."

Maurice de Broglie specialized in experimenting with the recently discovered X-rays and he repeatedly emphasized to his younger brother the dual aspects of particle and wave. Louis-Victor de Broglie also owes an intellectual debt to Marcel-Louis Brillouin (1854 – 1948) for his work on a vibrating particle in an elastic medium. However, that aside, it must be emphasized that the ideas of L. de Broglie were uniquely his own. In his dissertation for a Ph.D. presented to the University of Paris in 1924 he not only presented a very scholarly history of light but also the radical suggestion that all particles have a wave nature. In fact, he connected the wavelength (now known as the de Broglie wavelength) of a particle with its momentum. Here we have for the first time the dual personalities, wave and particle, linked in one equation.

There are at least two accounts by de Broglie of how he came to his result. "I was convinced that the wave-particle duality discovered by Einstein in his theory of light quanta was absolutely general and extended to all of the physical world, and it seemed certain to me, therefore that, the propagation of a wave is associated with the motion of a particle of any sort—photon, electron, proton, or any other."

Around 1962 he again recalled. "After the end of World War One, I gave a great deal of thought to the theory of quanta and to the wave-particle dualism. It was then that I had a sudden inspiration. Einstein's wave-particle dualism was an absolutely general phenomenon extending to all physical nature. The new concept also gave the first wave interpretation of the conditions of quantizing the momenta of atomic electrons." Here de Broglie was referring to Einstein's 1905 paper that explained the photo-electric effect. The quantizing of atomic electrons referred to by him comes about as follows. If an electron is treated as a wave in a circular orbit, then the wavelengths must fit exactly around the circumference. Imposing this condition yields precisely the Bohr radii.

In his 1929 Nobel Speech, de Broglie explained his ideas as follows. "On the one hand, the quantum theory of light cannot be considered satisfactory since it defines the energy of a light particle (photon) by the equation $E = hf$ containing the frequency f. Now a purely particle theory contains nothing that enables us to define a frequency; for this reason alone, therefore, we are compelled, in the case of light, to introduce the idea of a particle and that of frequency simultaneously. On the other hand, determination of the stable motion of electrons in the atom introduces integers, and up to this point the only phenomena involving integers in physics were those of interference and of normal modes of vibration. This fact suggested to me the idea that electrons too could not be considered simply as particles, but that frequency (wave properties) must be assigned to them also."

De Broglie's professor, Paul Langevin (1872 – 1946), had studied under J. J. Thomson and Rutherford. That de Broglie's ideas were very original, even radical is revealed by Langevin's confession about his student. "His ideas, of course, are nonsensical, but he develops them with such elegance and brilliance that I have accepted his thesis." Nevertheless, he sent a copy of the thesis to Einstein who recognized the significance of what de Broglie had achieved. In a letter to Langevin he wrote, *"Er hat eine Ecke des grossen Schleiers gelüftet."* (He has lifted a corner of the great veil.)

On de Broglie's oral exam, Jean-Baptiste Perrin (1870 – 1942) asked him if his waves could be observed. The young de Broglie answered that they should leave interference patterns just like X-rays if passed through crystals. Alexandre Dauvillier (1892 – 1979), to whom this experiment was proposed, was too busy working on developing television to undertake this experiment in 1924 and thus missed observing this fantastic result and earning the Nobel Prize.

In 1929 de Broglie's work was recognized with the Nobel Prize in Physics "for his discovery of the wave nature of electrons."

Several years later, at a dinner Count Louis de Broglie shared a table with P. A. M. Dirac and struggled all evening in English. At the end of the evening as de Broglie arose to leave he bade Dirac, *"Bon nuit."* Dirac replied in flawless French. Amazed, de Broglie asked, "Why did you not tell me that you speak French?" The response was typical Dirac, "You did not ask me."

There is another story, due to George Gamow (1904 – 1968) that throws a different light on this. When Gamow went to visit de Broglie at his magnificent castle, the door was opened by a very distinguished looking butler who, to Gamow's statement that he would like to see Professor de Broglie, replied "You mean the duke de Broglie." After admitting that he wanted to see the duke, Gamow was admitted. Once inside, the conversation was conducted entirely in French of which Gamow claimed to have but a poor command. A year later de Broglie came to London and delivered a lecture in very clear English.

At a later date when Gamow was again invited to give a lecture at the Sorbonne, he had failed to prepare his lecture in French since he had enjoyed his ocean voyage from America so much. Nevertheless, with his halting French the lecture seemed to go quite well and he was understood. Later when he apologized to de Broglie for not having corrected French notes, the latter replied, "It's a good thing you didn't. R. H. Fowler had come to give a lecture and, at his request, I had translated his notes into French. Unfortunately after the lecture some students came to me and questioned what language Professor Fowler had spoken since they had expected the lecture to be in English which they all thought they understood sufficiently well. I had to inform them that Professor Fowler had spoken in French."

The first observations of de Broglie's waves were made almost simultaneously in England by Sir George Paget Thomson (1892 – 1975), the son of Sir J. J. Thomson, and at the Research Laboratories of the American Telephone and Telegraph Company (later the Bell Labs) by Clinton Joseph Davisson (1881 – 1958) and Lester Halbert Germer (1896 – 1971). Thomson used an arrangement of crystals as in X-ray diffraction but instead of X-rays he used a beam of electrons of definite speed instead of X-rays to carry out precisely the experiment the young de Broglie had suggested in his thesis defense and observed a diffraction pattern. The wavelength that was obtained from the diffraction pattern agreed with that obtained from the de Broglie formula. G. P. Thomson and C. J. Davisson shared the Nobel Prize in Physics "for their experimental discovery of the diffraction of electrons by crystals."

Davisson had studied under Millikan at the University of Chicago until his scholarship ran out and his studies were interrupted due to finances. Millikan helped to get him appointed as an assistant in physics at Purdue until, again on Millikan's recommendation, he was appointed as part-time instructor of physics at Princeton under Jeans and Richardson on a fellowship that allowed him to complete his Ph.D. He was also instructor at the Carnegie Institute of Technology for six years and spent an interlude at the Cavendish Laboratory with J.J. Thomson.

The Davisson-Germer result was more serendipitous than G. P. Thomson's. Throughout 1922 C. J. Davisson and H. Kunsman had tried to utilize electrons to study atomic structure in the same way that Rutherford had used alpha particles. The results were rather disappointing and Kunsman left in 1923. These "disappointing" results had shown some maxima and minima. In Europe, Walther M. Elsasser (1904 – 1991) interpreted them as interference. Werner Heisenberg wrote to Wolfgang Pauli that this paper was important.

In the fall of 1924 Davisson started again, this time with the help of Germer. The year's delay and subsequent start-up had produced a crack in the vacuum system and exposed the polished nickel target to air causing some oxidation on the surface. To remove this oxygen, they heated the target for a prolonged time. This freak accident turned out to be a lucky break for, when on April 6, 1925 the equipment started working again, Davisson and Germer found a series of maxima and minima in the intensity of the scattered electrons. This strange behavior so puzzled them that they stopped the experiment, opened the vacuum tube and examined the nickel target. They found that the prolonged heating had caused the target to recrystallize in about ten larger crystals rather than a host of minute crystals.

They then realized that the crystalline structures and not the individual atoms were responsible for these maxima and minima. So, they set out to obtain a large single crystal for their target. A year later (1926) they were ready to try again, but the results were disappointing. Davisson left for a few months of a second honeymoon in England.

In the meantime three different versions of quantum mechanics, as we know it today, had been formulated in less than a year. First came the matrix mechanics of Werner Heisenberg, Max Born, and Pascual Jordan, soon followed by the wave mechanics of Erwin Schrödinger. This was followed by another formulation due to a young Englishman, P. A. M. Dirac. Thus, in England, Davisson was very much surprised to find physicists discussing his and Germer's early results as experimental confirmation of de Broglie's waves.

As Schrödinger's approach most naturally follows the discussion of de Broglie's waves, I present it next, even though matrix mechanics was developed before it.

The Davissons' second honeymoon was probably spoiled because Davisson now began to study Schrödinger's papers in detail. Finally back at Bell Labs he planned, together with Germer and a young electrical engineer C. Calbick, a series of more accurate experiments that finally on Jan. 6, 1927 resulted in a complete vindication of de Broglie's thesis.

In 1928, Davisson summarized the situation as follows, "We think we understand the regular reflection of light and X-rays—and we should understand the reflection of electrons as well if electrons were waves instead of particles. It is rather as if one were to see a rabbit climbing a tree, and were to say, 'Well, that is rather strange thing for a rabbit to be doing, but after all there is really nothing to get excited about. Cats climb trees—so that if the rabbit were only a cat, we would understand its behavior perfectly.'"

Chapter 10

Schrödinger Makes Waves

"But for the present these are just chimeras; it could be completely differ-ent." E. Schrödinger in a letter to H. A. Lorentz, June 1926.

Unlike Planck's original idea of a quarter of a century earlier, de Broglie's ideas had an almost immediate impact. The most lasting of these occurred in Zürich, Switzerland and the resultant tremors spread across the physics world.

In 1925, Zürich boasted two physics departments. One was at the University of Zürich, headed by Erwin Schrödinger (1887 – 1961). The second, more prestigious physics department was at the Eidgenössische Technische Hochschule or ETH, headed by Pieter Joseph Willem Debye (1884 – 1966). Since neither faculty was very large, they held their weekly colloquia together, alternating between the two institutions.

After one of these colloquia Debye suggested to Schrödinger that since he wasn't working on anything particularly important it would perhaps be good if he gave a talk on these new ideas of de Broglie's.

A few weeks later Schrödinger presented a talk that was a model of clarity in describing how de Broglie's ideas led to Bohr's rules of quantization. Debye was not impressed by these ideas and stated that this was a childish way of talking. After all, as a student, he had already learned from Sommerfeld that in order to deal with waves one needed a wave equation.

This remark may well be what triggered Schrödinger. For, although he later acknowledged Einstein for drawing the importance of de Broglie's work to his attention, he published his first paper on the now famous Schrödinger equation only a few months after giving this talk. Whatever the stimulus, Erwin Schrödinger was eminently prepared to tackle wave equations.

Schrödinger had been raised with a good classical education. His father owned a small, but quite successful, linoleum and oil cloth factory in Vienna. So, the family was quite well off and Erwin was tutored at home until age ten. His mother Georgine Emilie, née Bauer was half English and Erwin spoke both German and English at home. In 1898, he entered Vienna's Akademisches Gymnasium. He described his high school years as enjoyable. "I was a good student in all subjects, loved mathematics and physics, but also the strict logic of the ancient grammars, hated only memorising [sic] incidental dates and facts. Of the German poets, I loved especially the dramatists, but hated the pedantic dissection of their works."

Years later his friend and former Sommerfeld student, Paul Peter Ewald (1888 – 1985), decided to test Schrödinger, so he wrote him a letter in Latin. Schrödinger replied in classical Greek.

After graduation from the gymnasium in 1906 he began his studies at the University of Vienna under Fritz Hasenöhrl (1874 –1915) who had taken over the chair that became empty with the untimely death of his teacher, Ludwig Boltzmann. Hasenöhrl's lectures on theoretical physics had a profound and long lasting influence on Schrödinger. He graduated in 1910 and got his *Habilitation* in 1914. In between he volunteered for military service in an artillery unit. So, when World War One broke out he was ordered to an artillery unit on the Italian border. His duties must not have been too onerous for he soon published some research and continued to do so after he was moved to Hungary. He was again moved back to Italy where he was cited for outstanding service commanding a battery during battle. In 1917, Schrödinger was called back to Vienna to teach a course in meteorology. He continued to do some fine, but not outstanding research on various aspects of physics.

In 1920, Schrödinger was offered the position of Extraordinary Professor at Vienna and in the same year married Anny Bertel to whom he had been engaged for a year. At that time Anny was working as a secretary in Vienna on a monthly salary which was more than Schrödinger's annual income. Instead of the position in Vienna, he accepted a position as assistant in Jena. This paid well enough for him to marry.

Erwin and Anny's marriage was full of stress. They had an open marriage and he had numerous lovers, as did Anny. For a while the mathematician and Erwin's good friend, Hermann Weyl, was Anny's lover. In 1933, the Schrödingers together with several other scientists spent the summer in South Tyrol. Here Erwin had an affair with Hilde March who bore him a daughter, Ruth Braunizer (née March). It is probably the birth of his daughter that caused Schrödinger to turn down an offer from Princeton University in 1934. He wanted to raise his daughter with both Anny and Hilde. This was not acceptable in the less liberated Princeton community. Accordingly Schrödinger accepted an offer from the University of Berlin with the condition that Hilde's husband be hired as his assistant.

When Robert Oppenheimer, the director of the Princeton Institute for Advanced Studies, visited the University of Alberta to open the new physics building someone asked him why Schrödinger had never visited the Institute in Princeton. His response was that he had invited Schrödinger and had received the reply that, "As you know, I not only have a wife but also a mistress and I would like you to assure me that on my visit they will both receive an equally cordial welcome." To this Oppenheimer had replied that he could assure Schrödinger that he would certainly treat them both in a most cordial manner, but he could not vouch for other people's behavior. Schrödinger then wrote back, "In that case I can't come."

Schrödinger's sexual exploits are well known. His first acknowledged mistress was the seventeen year old Itha Junger. Also, much later in life in trying to seduce a young woman in Ireland he successfully used the approach, "If you can't love me as a lover because I am a married man, just love me as a father." In fact, in Ireland, he fathered two children with two different women. He seems to have been what today is called a "male chauvinist" as displayed by his statement about his mistresses. "Poor things, they have provided for my life's happiness and their own distress. Such is life."

It must have somewhat shocked the citizens of Catholic Dublin to see Schrödinger walking down the streets of their city with his wife on one arm and his mistress on the other. It had certainly shocked his colleagues in Cambridge when he spent a year there in 1934.

Throughout his life Schrödinger was somewhat of a bohemian. When, in the 1920s, he was invited to that most conservative of German institutions, the University of Berlin to give a seminar, he showed up in knickerbockers while his audience wore stiff white collars. Further proof of Schrödinger's tendency to disdain conventions is related by Dirac in Robert L. Weber's *Pioneers of Science: Nobel Prize Winners in Physics*. He tells that normally when Schrödinger attended a conference, he arrived with all his luggage in a rucksack and walked from the station to the hotel. His attire was so casual that he might easily be mistaken for a tramp. This meant that typically quite a discussion, at the reception desk of the hotel, ensued before he could get a room. His irreverence to fame and authority also extended to his colleagues and so he was also the one who first referred to Niels Bohr as the "Great Dane".

In 1925, Schrödinger took his Christmas vacation at Arosa in Switzerland and stayed there till January. His marriage was strained at the time and he wrote to an old girl friend, who remains nameless, to join him. Schrödinger believed in the inspirational powers of sex and his friend, whoever she was, seems to have inspired him. As his close friend, Hermann Weyl, later wrote, "Schrödinger did his great work during a late erotic outburst in his life."

After his vacation in Arosa, Schrödinger was full of energy. On his first attempt to find a wave equation he started with a relativistic formulation of the old problem of the electron in a hydrogen atom. After completing the computation he found a spectrum very similar to that correctly given by the Bohr-Sommerfeld rules, but sufficiently different to be obviously wrong. So, he rejected his result without publishing it. As we now know, Schrödinger's formulation was correct but describes mesons, not electrons. The correct relativistic equation for electrons was found two years later by P. A. M. Dirac.

According to Dirac. De Broglie's ideas were for a free electron by itself, and Schrödinger extended them to apply to an electron moving in an electromagnetic field. As soon as he got his general equation, he applied it to the hydrogen atom. The result that he obtained was not in agreement with experiment, because Schrödinger did not know at that time about the spin of the electron. He was extremely disappointed by this failure. He told me about it many years later. He believed that the whole idea of his wave equation was wrong. He was terribly dejected, and he abandoned it altogether. Then it was some months later that he recovered from his depression sufficiently to go back to this work, look over it again, and to see that if he did it in a nonrelativistic approximation, so far as a nonrelativistic system was concerned, his theory was in agreement with observation. He published his equation then as a nonrelativistic equation.

You may wonder how it appears that Schrödinger's early papers were all nonrelativistic, although they were inspired by de Broglie waves and the de Broglie waves were built up from relativistic ideas. It was in this indirect way that it came about. Schrödinger lacked courage to publish an equation that gave results in disagreement with observation. He should have had that courage; he would then have published a second-order equation ... an equation that was later to be known as the Klein and Gordon equation, although it was discovered by Schrödinger before Oskar Klein and

Walter Gordon and was the first wave equation that he worked with. But Schrödinger would only publish something that was not in direct disagreement with observation. People were rather timid in those days, I suppose, and it was left to Klein and Gordon to publish an equation which is now accepted as the correct equation for a charged particle without spin.

At any rate, in 1926 mesons were totally unknown and Schrödinger disappointedly set his work aside. As he later told Dirac, "I immediately applied my method to the motion of the electron in the hydrogen atom, properly taking into account the formulas of the relativity theory for such an electron. ... The calculated results did not agree with the observational data ... I was deeply disappointed, decided that the method was unsuitable, and dropped it."

Incidentally, the relativistic wave equation that Schrödinger discovered is now known as the Klein-Gordon equation after two physicists who, along with numerous others, rediscovered it later. This led Pauli to call it the "equation with many fathers".

Fortunately, Schrödinger's disappointment did not last. A couple of months later, after returning to Zürich, he returned to his work and attempted a nonrelativistic formulation. About this time he wrote, "A moving particle is nothing else but the foam on the wave radiation forming the matter of the world." The result was the now celebrated Schrödinger equation on which most of modern electronic technology and solid state physics as well as almost all of quantum chemistry is based. In the introduction to his paper—published in *Annalen der Physik* in 1926—he stated, "I wish to consider the hydrogen atom and show that the customary quantum conditions can be replaced by another postulate, in which the notion of 'whole numbers' merely as such is not introduced. Rather when integralness does appear it arises in the same natural way as it does in the case of the *node-numbers*, of a vibrating string. The new conception is capable of generalization, and strikes, I believe, very deeply at the true nature of the quantum rules."

The hydrogen atom had by now become the touchstone for all quantum theories. The results of Schrödinger's calculations fit beautifully. Nature had once more given up one of her secrets.

How had Schrödinger attacked this problem? In a most ingenious fashion he had turned the problem inside out. Thus, recalling that William R. Hamilton (1805 – 1865) had shown that ordinary mechanics was mathematically equivalent to the motion of waves in a ray optics approximation, Schrödinger asked himself, the following question. "What must the wave equation be, such that for a free particle it yields the de Broglie wavelength as well as ordinary mechanics in the ray approximation?" The rest is history. Well, not quite.

What kind of wave was this? What was it that was vibrating? It was not electromagnetic fields, nor was it matter. And it certainly wasn't a non-existent ether. What was it? One way to hide one's ignorance about something is to label it. Thus, Schrödinger called his waves psi (ψ) or matter waves. This lack of meaning for Schrödinger's waves led Walter Hückel (1895 –1973), a collaborator of Debye, to write the following poem, which, has been translated by Felix Bloch (1905 – 1983).

> Erwin with his psi can do
> Calculations quite a few
> But one thing has not been seen
> Just what does psi really mean.

Schrödinger believed that these waves described the charge density of the electrons but, as shown by Heisenberg and others, this view was not tenable. Soon afterwards, Max Born, who had already played a decisive role in the formulation of matrix mechanics, gave what is now the accepted interpretation.

At any rate for many physicists, including Planck, who were abhorred by the discontinuities in Planck's original formulation of quanta, Schrödinger's work seemed to be a godsend. Planck described it as "epoch-making work". Einstein also wrote, "... the idea of your work springs from true genius ... " He continued, "I am convinced that you have made a decisive advance with your formulation of the quantum condition." Ehrenfest was even more lavish in his praise. "I am simply fascinated by your [wave equation] theory and the wonderful new viewpoint it brings. Every day for the past two weeks our little group has been standing for hours at a time in front of the blackboard in order to train itself in all the splendid ramifications." Quantum theory seemed understandable again in terms of the familiar concepts of waves and charge densities. In 1933, Erwin Schrödinger shared the Nobel Prize in physics with Paul Adrien Maurice Dirac "for the discovery of new productive forms of atomic theory".

This joy, that quanta could again be understood in terms of classical concepts, was short-lived. For, while the wave equation remained as a permanent part of physics, the simple interpretation did not. On the other hand, it was clear that Schrödinger's approach far excelled the Bohr-Sommerfeld quantization rules. Not only could Schrödinger handle non-periodic motions but he could also answer more complicated questions that were beyond the pale of the old quantum theory. Thus, he could determine not just the frequencies but also the intensities of the spectral lines of hydrogen, a feat that was possible with the Bohr-Sommerfeld rules only after some additional ad hoc modifications. The results also agreed splendidly with experiment.

The difficulty with Schrödinger's interpretation of his waves was brought to a head by a young genius, Werner Heisenberg, who only a few months earlier had been instrumental in formulating a completely different version of quantum mechanics. This so-called matrix mechanics contained all the original discontinuities of Planck's quanta with a vengeance and forced a radical reinterpretation of the Schrödinger waves.

Schrödinger also made important contributions to biology. In 1943 in his book, *What is Life?* he wrote, "The chromosome is a message written in code." This book, although a tremendous success and translated into many languages with an estimated sale of about 100,000, netted very little money for the author. It was also responsible for converting James Dewey Watson (1928 –) and Francis Crick (1916 – 2004), of double helix fame, from physics to biology. This is what Watson later stated. "From the moment I read Schrödinger's *What is Life?* I became polarized towards finding out the secret of the gene." Francis Crick had this to say about theory. "Any theory that agrees with all the facts must be wrong, because some of the facts must be wrong."

After the Nazis took over in Germany, Schrödinger could no longer enjoy his idyllic life in Berlin in the company of Planck, Einstein, and von Laue. He was unable to tolerate what was going on. One day he saw storm troopers in action, in front of Wertheim's, the largest Jewish department store in Berlin, roughing up people who wanted to enter the store. He tried to interfere and gave one of the storm troopers a piece of his mind. The other storm troopers were ready to attack him and might have beaten him to death. Fortunately, Friedrich Möglich, an assistant to Max von Laue, was also wearing a Nazi insignia, recognized him and led him away.

While in Rome, the Prime Minister of Ireland, Eamon de Valera a mathematician with an interest in physics, met Schrödinger and convinced him to become the star in the newly founded Dublin Institute for Advanced Studies. Schrödinger had had enough of Nazi Germany and left.

In 1940 when Ireland, so as to not offend the German government, imposed increasing censorship, Schrödinger grew more pessimistic about an allied victory. At this time he gave an interview to the *Irish Press*. His final statement was, "There is no worldly truth but mathematical truth. In politics, diplomacy, history, truth changes from day to day, and people get different concepts of right. But mathematics never lie."

When Schrödinger heard of Hiroshima and the atom bomb, he was devastated. He wrote a letter to his close friend Hermann Weyl. "I find development of things on this planet so desperate that I close my eyes and don't look around. ... The dangerous enemy is the State. The abscess of fascism has been cut out, but the idea lives on in its sworn enemies. ... I shudder at the thought that it can go so far with us, but it already has gone much too far. For example, the atom bomb."

He then proceeded to upset the secretive military establishments of Britain and the USA by deriving, solving and publishing the so-called Peierls equation for the critical mass for an atom bomb. This formula is crucial for the design of an atom bomb. Of course he could get away with it since Ireland was not subject to the prohibitions pertaining to nuclear secrets.

There are several stories from the time that Schrödinger spent in Dublin. When Herbert S. (Bert) Green (1920 – 1999) was a visiting professor at Dublin, he arranged the seminars. Having just read a paper in the *Proceedings of the Royal Society* with the title, "The Statistical Interpretation of Quantum Mechanics" he thought this might make a good topic for the seminar. Schrödinger disagreed. "Oh yes, I refereed and rejected this paper because there is no statistical interpretation of quantum mechanics."

Sometime soon after the war an Austrian physicist, who had served in the military and retained much of that stiffness of manner, came to Dublin. His excessive formality and deference to Schrödinger began to annoy the latter. Thus, when this gentleman approached Schrödinger and addressing him in the manner of a superior officer asked him if he would mind reading his paper to give him permission to publish it, Schrödinger simply said, "Oh yes" and walked away. At the next encounter this gentleman again asked Schrödinger if he might be able to look at his paper. The answer was, "Tomorrow." The next day Schrödinger was not at the Institute. Still patient, this physicist again approached Schrödinger saying that he would really like him to read his paper and asked him when this might be possible. Schrödinger replied that he was busy for the next little while, but that if he wanted, he could bring the paper to his home that evening.

That evening, at his home, Schrödinger made some polite conversation and the physicist did not want to change topics to raise the matter of his paper. Soon thereafter Schrödinger got up and left. The gentleman waited patiently for 5, 10, 15 minutes, for half an hour, for an hour. No Schrödinger. So, he got up to look for him, but Schrödinger was nowhere to be found since he had gone to bed. Incidentally, they later became quite friendly and even published together.

One morning at the Dublin Institute, both Schrödinger and the cleaning maid tried to simultaneously quit their jobs. He because she destroyed some of his work, and she because he would not let her do her job. Apparently Schrödinger had thrown some of his calculations into the wastebasket the previous evening and the cleaning maid had promptly emptied this basket. The director John Leighton Synge (1897 – 1995) solved the situation by stating that henceforth Schrödinger would have to empty his wastepaper basket himself.

Synge was also famous as a relativist and mathematical physicist in his own right. In the introduction to his classic book *Relativity : The General Theory* he wrote, "Splitting hairs in an ivory tower is not to everyone's liking and no doubt many a relativist looks forward to the day when governments will ask his advice on important questions."

On one occasion, at the Dublin Institute, a speaker kept referring to "Synge's Lemma". After several such references, Synge inquired as to what exactly Synge's Lemma was. On hearing the explanation he corrected, "That should be Synge's Theorem." For those who don't know, a lemma is sort of a mini-theorem used to help prove a theorem.

There is one more Dublin story. A physicist named Janossi was like a broken record every morning at tea. "Communism was the worst possible system." One morning he was uncharacteristically silent. The same the next day, and the day after. On the fourth day he commenced an attack on capitalism and continued this assault for the next week. Then, he disappeared. His colleagues later discovered that he had been appointed director of the institute in Budapest.

Schrödinger remained in Dublin until 1955 when he returned to Vienna so that he or his wife would not have to live on a pension of a mere £50.00 a month, but could instead receive the pension of an Austrian professor. Eventually he retired to the village of Alpbach in Tyrol where he died in 1961.

When I visited the cemetery to view his grave, which was supposed to have a tombstone with his equation inscribed on it, I was unable to locate it. At the inn across the street from the churchyard the barmaid informed me that, "He lies right over there against that wall." I found the grave. It bore a simple wrought iron cross with his name and dates. I returned to the inn and asked about his equation. Again I was told, "Oh, the priest didn't like those cabalistic symbols." Apparently there was also some difficulty over burying him in the churchyard since Schrödinger had not been a practicing Catholic. The village priest was finally convinced to allow a spot, near the wall away from the good Catholics, after he was informed that Schrödinger had been a member in good standing of the Papal Academy.

On another occasion, about 1985, while skiing near Alpbach I talked to a young man from that village as we rode up on the lift. I mentioned to him that he must be proud that Schrödinger, the man honored by his government to have his face displayed on the 1,000 Schilling (about $80.00) note, was buried in his village. The young man looked at me with a puzzled expression, "Who was he anyway?"

Chapter 11

Boys' Physics and Quantum Jumping

"The present paper seeks to establish a basis for theoretical quantum mechanics founded exclusively upon relationships between quantities which in principle are observable." Werner Heisenberg, Zeitschrift für Physik, 1925.

The years 1925 – 1927 became known among physicists as *Knabenphysik* (boys' physics) since all the creators of quantum mechanics, with the exception of Born and Schrödinger, were in their early twenties. These young men were: Heisenberg, Dirac, Jordan, and Pauli.

Almost a year before Schrödinger developed his equation, the young twenty-five year old, Werner Heisenberg (1901 – 1976), originated what came to be known as Matrix Mechanics.

Young Heisenberg came from the upper middle class. His father August Heisenberg taught Greek at the University of Munich. Werner's maternal grandfather, Nikolaus Wecklein, was the headmaster of the Maximilian Gymnasium in Munich. After Werner's family moved to Munich from Würzburg and the young boy finished elementary school he entered the gymnasium where his grandfather was headmaster.

The outbreak of World War One disrupted Heisenberg's schooling. Troops took over the school and classes were held in various locations so that young Werner did quite a bit of independent study. His progress in mathematics was so good that in 1917 while still in gymnasium he tutored a university student in calculus. About this time he joined the *Pfadfinder*, a sort of boy scout organization. His friendship with the other young men he met in this organization played an important role in this part of his life. He wrote of this period. "My first two years at Munich University were spent in two quite different worlds: among my friends of the youth movement and in the abstract realm of theoretical physics. Both worlds were so filled with intense activity that I was often in a state of great agitation, the more so as I found it rather difficult to shuttle between the two." The war caused great food shortages and Werner joined a volunteer organization to work on the farms to replace the men on the front and help bring in the harvest.

When the war ended, Germany was in turmoil. Various factions attempted to take power. Young Werner ran messages through the military lines to the north of Munich for the forces fighting the Bavarian Soviet forces. This was very dangerous work, but as he himself stated, "I was a boy of seventeen and I considered it a kind of adventure. It was like playing cops and robbers."

In 1920, Werner won a scholarship to the University of Munich. Originally he had thought of either studying mathematics or becoming a concert pianist. He did not view himself good enough to make a career in music and there are two stories why he chose physics over mathematics.

Heisenberg had read Hermann Weyl's book *Space, Time, Matter* and both excited and disturbed by the abstract mathematical arguments in that book decided to study mathematics. To this end he wanted, as a young student at the University of Munich, to attend the seminar of Professor Ferdinand von Lindemann (1852 – 1939), famous for solving the ancient problem of squaring the circle. Lindemann's proof consisted in showing that the number π is transcedental. This means that π is not the solution of any polynomial equation with rational coefficients. From this result it followed that it is impossible, by using only a compass and straightedge, to construct a square with the same area as a circle.

Heisenberg's father arranged an interview for him with the famous professor, so that he could obtain permission to attend Professor Lindemann's seminar. According to Heisenberg the interview was a disaster. When he entered the great professor's gloomy office, furnished in a formal, old-fashioned style, he almost immediately felt a sense of oppression. A little black dog cowered on the desk in front of the professor and glared at him with open hostility. The young Heisenberg was so flustered that he began to stammer so that his request began to sound immodest even to him. Then the little dog began to bark horribly and Professor Lindemann's attempts to calm the mutt were to no avail. As a consequence the interview turned into a shouting match over the barking of the dog. When Lindemann asked what he had read, Heisenberg mentioned Weyl's book. Over the incessant yapping of the dog Lindemann shouted, "In that case you are completely lost to mathematics." Another fact that may have contributed to Lindemann's impatience was that he was only two years from retirement and had agreed to interview the young man only as a courtesy to his father.

So, since Heisenberg was "unfit" for mathematics, physics benefited and physicists are duly grateful to Lindemann.

With regard to Weyl's book there is another amusing incident. Later in life, when Heisenberg was an Extraordinary (Associate) Professor, his first graduate student was Felix Bloch (1905 – 1983). One day on a walk, they started to discuss the concepts of space and time. Bloch had also just finished reading Weyl's *Space Time Matter* and was still very much under the influence of this scholarly work. Thus, he declared that he now understood that space was simply the "field of affine transformations". Heisenberg paused and looked at him, "Nonsense, space is blue and birds fly through it."

There is a second, probably more credible, version of why Heisenberg chose physics over mathematics. As a nineteen year old student at the University of Munich, he went to the University of Göttingen to hear lectures by Niels Bohr. This was the first occasion since the end of World War One that an important foreign visitor had come to Germany. It was an indication that Germany was being accepted back into the community of nations. Physicists and their students from all over Germany attended. The whole thing took on the semblance of a festival so that the lectures were jokingly referred to as the "Bohr Festspiele" (Bohr festival). This was an allusion to the Händel festival which was occurring at the same time.

At this meeting, Bohr expounded on his latest theories of atomic structure. The young Heisenberg in the audience did not hesitate to ask questions when Bohr's explanations were less than clear. This so impressed Bohr that after the lecture he invited the young man to go for a walk with him to drink beer, eat a snack, talk about physics, and "have a good time". This excursion which lasted several hours impressed Bohr with the young man's talents. Heisenberg in turn was impressed with the Danish physicist's way of attacking problems by first trying for a way to match ideas with experimental results before trying to give a deep mathematical analysis. Also Bohr acknowledged that he did not know the answers to many of Heisenberg's questions. Thus, these problems became alive to the young man.

A sequel to this is that, on the evening after their walk at the banquet, two German policemen in uniform came to "arrest" Bohr for "kidnapping small children". The policemen were two graduate students playing a prank.

While studying with Sommerfeld, Heisenberg was fortunate to have a senior classmate, Wolfgang Pauli (1900 – 1958), whose influence he acknowledged with the words, "Pauli had a very strong influence on me. I mean, Pauli was simply a very strong personality ... He was extremely critical. I don't know how often he told me, 'You are a complete fool' and so on. That helped a lot." Later in life he claimed that "Pauli is such a strong critic that if he cannot produce a counter-argument (counter-example), the result must be right and should be published."

Sommerfeld had this to say about Pauli. "I can't teach him anything; at my suggestion he is writing a summary of Einstein's relativity theory." Actually Sommerfeld had been asked to write this article for the *Encyclopädie der mathematischen Wissenschaften* and had passed the work on to Pauli. This book, written by a nineteen-year-old, appeared in 1921 and contained the prophetic statement, "Perhaps the theorem on the equivalence of mass and energy can be checked at some future date by observations on the stability of nuclei." Today it is still considered one of the better books on Relativity.

One of the early problems that Heisenberg worked on while studying with Sommerfeld was the so-called anomalous Zeeman effect.[1] The Zeeman effect is what physicists call the splitting of spectral lines of an atom when it is put in a magnetic field. The word "anomalous" here refers to that part of the effect that could not be explained by classical theory.

When in 1920 Heisenberg reported to Sommerfeld that he could explain some of the anomalous Zeeman lines by using half-integer quantum numbers the latter exclaimed, "That is absolutely impossible! The only fact we know about quantum theory is that we have integral numbers and not half integers." But in 1921 when Alfred Landé (1888 – 1975) came to Munich and proposed a set of rules that involved half integers, Sommerfeld noted to him, "Your new representation agrees well with what has been found by one of my students (in the first semester), but which has *not* been published."

Pauli was also critical of these half-integer quantum numbers. Heisenberg justified his approach with, "Success sanctifies the means." In a letter to Bohr in February 1924 Pauli, though critical of Heisenberg's approach to physics, wrote of him, "But if I talk to him, he strikes me as alright, and I see that he has all sorts of new arguments—at least in his heart. I therefore think of him—aside from the fact that he is also personally a very

---

[1]The Zeeman effect is discussed in more detail in Chapter 15.

nice fellow—as very thoughtful, even a genius, and I think that he will once again greatly advance science."

After Heisenberg completed his studies with Sommerfeld he made a pilgrimage to the Mecca of quantum physics, to Niels Bohr's institute in Copenhagen. Here Heisenberg was soon deeply engrossed in the problems of reconciling Bohr's quantum with the numerous inconsistencies that were becoming more and more obvious.

In 1923 an American, John Slater (1900 – 1976), arrived in Copenhagen with the marvelous idea that the transitions in a Bohr atom were not instantaneous, but lasted for a short time related to the width of the photons emitted. He also came up with the idea that all transitions were present at all times and did not carry energy, but determined a probability for finding a photon at a given place. Bohr and Hendrik Anthony Kramers (1894 – 1952) received his idea with enthusiasm. They modified this approach and the result was the so-called BKS (Bohr-Kramers-Slater) theory which used a "virtual" or "ghost" radiation field that carried no energy or momentum to maintain a wave theory in interaction with matter. Einstein found this most repugnant since conservation laws and causality were now purely statistical.

In a letter in February 1924, Pauli wrote to Bohr regarding this virtual oscillator theory of radiation. The sarcasm is obvious. "I have tried on the basis of the definition of the two words (*kommunisieren, virtuel*) to guess what your work is really about. But it is not easy. In any case, it is very interesting to me and if I can be of any help with the grammar I shall gladly oblige." In a letter to Heisenberg, Pauli also complained about the "virtualization" of physics.

Soon after this, Pauli was able to demolish the BKS theory by checking carefully its relativistic implications. In a letter to Sommerfeld in 1925 Pauli wrote, "One now has the strong impression with all models that we are speaking a language that is not sufficiently adequate to the simplicity and beauty of the quantum world."

Later in 1925, John Hasbrouck van Vleck (1899 – 1980) who shared the 1977 Nobel Prize in Physics with Philip W. Anderson and Sir Nevill F. Mott "for their fundamental theoretical investigations of the electronic structure of magnetic and disordered systems", stated regarding the death of the BKS theory, "Modern physics is certainly passing through contortions in its attempt to explain the simultaneous appearance of classical and quantum phenomena; but it is not surprising that paradoxical theories are required to explain paradoxical phenomena."

Bohr's reaction to this, in a letter to Werner Heisenberg, was depressing, "There is much to report, for the most part negative, for I look at many things even more doubtfully than at the time that you were here. With Pauli's help I endure the torture in order to get used to the mysticism of nature. I am preparing myself for all eventualities."

Earlier, while in Copenhagen and all this turmoil was occurring, Heisenberg had written, "If I were there I would, as in the case of the Zeeman effect, plead for a formal dualistic theory: everything must be describable, both in terms of the wave theory and in terms of light quanta."

In the spring of 1925 Heisenberg went to Göttingen to work for a few weeks as Max Born's assistant. This is how Born described the young man. "He looked just like a simple farm boy with his short light hair, clear light eyes and a radiant facial expression. His incredible quickness and accuracy of comprehension enabled him to continually achieve a tremendous amount of work without great effort."

When Born asked him about his future plans Heisenberg replied, "I don't have to decide that! Sommerfeld decides that!" That same year Heisenberg wrote to Pauli, "Unfortunately my own private philosophy is far and away not so clear, but rather a mishmash of all possible moral and esthetic calculation rules through which I myself often cannot find my way." In another letter to Pauli he wrote, "Our world is so-to-say the simplest of all possible worlds! But all that is music for the future and prior to that it is still necessary to do a lot of mathematics."

In reviewing his student days Heisenberg wrote later in life, "I learned optimism from Sommerfeld, mathematics at Göttingen, and physics from Bohr."

Near the end of May of that year, Heisenberg fell ill with a severe attack of hay fever and Born granted him a two-week leave. This he decided to spend on Heligoland, which is but a barren rock jutting out of the North Sea.

Not only did the recovery from his hay fever make swift progress here, but without distractions so did his physics. He soon settled on the Bohr orbits as his main stumbling block. Whereas Bohr had simply decreed that these orbits were stable and electrons did not radiate while in them, he had not really explained the stability. Even worse, none of these orbits could be observed; only the radial difference between two orbits could. Thus, Heisenberg asked himself whether in a physical theory it made sense to talk about something that couldn't be observed? After all, wasn't physics supposed to be an experimental science based on observation?

Considerations like this led the young man to borrow a principle that Einstein had employed. It seemed that to create the special theory of relativity, Einstein had eliminated such unobservable quantities as the ether and absolute simultaneity and had replaced them with observables such as the "proper time" of an individual. Similarly, Heisenberg adopted as a general guiding principle the idea that a physical theory must contain only observable quantities. Later, when Einstein began to question quantum mechanics and when Heisenberg got into a debate with him, the young man defended himself. "But, I only used the same principle as you." To this Einstein replied, "I may have used such a principle once upon a time, but it is wrong. It is like a joke which may be fine once, but is usually not good the second time."

In a very general sense Einstein was indeed right. A theory need not contain only observable quantities; the theory itself need only dictate which quantities are observable. This it turns out is, in fact, the case in quantum mechanics. Happily however, in spite of a questionable principle, Heisenberg's approach was inspired and the "joke" was still superb when told a second time. Instead of sticking to the unobservable Bohr orbits, Heisenberg compiled tables of the observable changes (differences) in the radii and momenta between different Bohr orbits. His objective was to find a method of computing directly with these tables of observable quantities instead of with the Bohr orbits themselves. As a check on the correctness of his calculations he had to obtain conservation of energy since all quantum processes, so far observed, obeyed this law.

In his memoirs Heisenberg recalls how he came up with matrix mechanics. One evening he started to compute the energies for a simple oscillator to see if they were conserved and found the results coming out correctly. Getting excited, he started to make mistakes and it was well past midnight by the time he completed his computations and knew that they worked. Energy was indeed conserved and his computations involved only quantities that were observable. He had worked only with his tables of observable changes in radii and changes in momenta between different Bohr orbits. The results were intoxicating. In his own words, "I was far too excited to sleep and so, as a new day dawned, I made for the southern tip of the island where I had been longing to climb a rock jutting out into the sea. I did so now without too much trouble and waited for the sun to rise."

What was it that this young genius had discovered? Instead of the usual orbits of classical physics, he now had tables of numbers, tables infinite in size and that had to be "multiplied" in a most peculiar manner. So, that if one had two tables of numbers labeled A and B, then $A \times B$ did not equal $B \times A$ with this type of multiplication. The only obvious advantage was that all the numbers in these tables corresponded to observable quantities.

The fact that $A \times B$ does not equal $B \times A$ is not really that strange when one realizes that each table of numbers, A or B, the entries correspond to a measurement or operation and their relative placement $A \times B$ or $B \times A$ tells which operation is done first. For example if A corresponds to putting on socks and B to putting on shoes then the result of A first and B second is very different from B first and A second.

After he returned to Göttingen, Heisenberg showed his work to Max Born (1882 – 1970) who immediately recognized its importance and sent it off to the editor of *Zeitschrift für Physik*.[2] This paper written in July 1925 and published in September 1925 was the beginning of Matrix Mechanics. Heisenberg used to refer to this first paper as "the great saw". When asked what that meant, his answer was, "It is the tool to saw off the limb on which the old physical theory rests."

Meanwhile, Born was disturbed by a vague familiarity in the unusual multiplication rules that Heisenberg had found for his tables of observable numbers. Then, he remembered that as a student in an abstract algebra course he had learned to multiply tables of numbers in just such a way. Born later recalled, "I began to ponder about his symbol multiplication, and was soon so involved in it that I thought the whole day and could hardly sleep at night ... And one morning ... I suddenly saw the light: Heisenberg's symbolic multiplication was nothing but the matrix calculus, well known to me since my student days." Heisenberg had rediscovered these abstract mathematical operations on the basis of sound physical intuition. Incidentally, Hamilton whose name has cropped up time and again during our brief history, is credited as one of the inventors of matrix algebra. The next step planned by Born was to distill the essential ideas in Heisenberg's work.

Wolfgang Pauli immediately accepted Heisenberg's work and wrote to Kramers, "On the whole I am now close to Heisenberg in my scientific opinions, and that our opinions agree in everything as much as is in general possible for two independently thinking men."

Only a few months earlier Pauli had been so discouraged by the inconsistencies and lack of insight in the Old Quantum Theory that, in a letter to Heisenberg, he had written that he wished he had become a cob-

---

[2]A translation of the abstract to this paper is at the top of this chapter.

bler or a tailor rather than to have studied physics. Pauli's encouragement greatly increased Heisenberg's confidence. This was important since there was strong opposition to this "abstract" algebra of matrices.

Heisenberg's work achieved immediate recognition and in 1927 at the age of 26 he was appointed Professor of Physics at the University of Leipzig.

Nine years later at the age of 35, Heisenberg was at a friend's house playing piano in a concert. There he met the 22-year-old Elisabeth Schumacher. She was not only beautiful, but also an accomplished musician, something that immediately resonated with Heisenberg. Less than three months later on April 29, 1937 they married. Accordingly, Heisenberg asked to delay his appointment in Munich until August rather than March. Music continued to remain an important part of their lives as they raised seven children.

During World War Two Heisenberg, unlike Schrödinger, remained in Germany. In his memoirs he stated that he contemplated leaving but as he looked out over the countryside of his beloved Bavaria he knew that he could not leave. This may not have been his only reason. In a letter to *Physics Today*, May 1991, Max Dresden wrote that in August 1939, the Laport family held a party at their house in Ann Arbor, Michigan for the visiting Heisenberg. A lively discussion about Nazi Germany took place. Heisenberg stated that he believed that with his prestige, he could influence and guide the German government. Enrico Fermi, who was also present, believed no such thing. "These people have no principles; they will kill anybody who might be a threat—and they won't think twice about it. You only have the influence they grant you." It may have been Heisenberg's belief that he could exercise such influence, as well as his often-expressed love for his country that persuaded him to stay. Although he believed that his country would be victorious he also stated that someone will "have to be there to pick up the pieces after it is all over".

That Heisenberg had no love for the Nazi's racist ideas was clear from his refusal to join the Nazi party and his continuing to teach relativity as before. This led the two staunch anti-Semites Lenard and Stark to attack him as a "white Jew". It was only his mother's connection with Himmler's mother and the intervention by the *Reichsführer-SS* that forced the Nazi rag *Der Stürmer* to tone down its vitriolic rhetoric and saved Heisenberg's position as professor. Even so, when Sommerfeld retired, Heisenberg, in spite of Sommerfeld's and other eminent physicists' recommendations, did not get the chair at the University of Munich that he wanted in order to be close to his parents.

Some further evidence of Heisenberg's attitude can be derived from his writings. In the fall of 1935 he wrote to his mother, "But I must be satisfied to oversee in the small field of science the values that must become important for the future. That is, in this general chaos, the only clear thing that is left for me to do. The world out there is really ugly, but the work is beautiful." During the war, when he was isolated and as he was turning 40, he again wrote, "He who has dedicated his life to the task of going after the individual connections of nature will be confronted over and over again with the question of how those individual connections fit harmoniously into the whole, other than the whole presented to us by the day to day life or the world." He also wrote, "It is more important to treat others humanely than to fulfill any sort of professional, national, or political duties."

One of the events during the war, that has been the subject of great controversy, numerous historical papers, and even a play *Copenhagen* by the British playwright Michael Frayn, was a visit by Werner Heisenberg in 1942 to Niels Bohr. They met secretly behind the Carlsberg Brewery in Copenhagen since the German forces were occupying Denmark and Bohr did not want to be seen with Heisenberg and Heisenberg did not want to be seen with Bohr. After the war Heisenberg claimed he went there to tell Bohr not to worry about the German A-bomb project and to ask him to convince the allies not to produce one. Bohr's version was different. Years later, Don Perkins asked Sir Rudolf Peierls, who knew them both well, which one was right. Peierls replied, "They're both wrong. Heisenberg only remembers what he wants to and Bohr could not possible have heard what Heisenberg said since he never stops talking."

Many years after the war when Bohr visited the Soviet Union, Elevter L. Andronikashvili (1910 – 1989) asked him about Heisenberg's staying in Nazi Germany and working on the atomic bomb project. Bohr thought for some time before answering. "Well, Aage[3], Professor Rosenfeld and I have discussed the question among ourselves and come to the conclusion that none of us could make any accusation against Professor Heisenberg. We then checked our conclusion with scientists who visited us from other countries. No, Professor Heisenberg is in the clear."

After World War Two, Phase III of the ALSOS (Army Logistics Support to Other Services) Mission of the Manhattan Project rounded up all the major German physicists that might have been involved with nuclear bomb research. Heisenberg was one of them. The American physicist in charge was Samuel Goudsmit (1902 – 1978) whose parents had died in one of the

---

[3]Aage Bohr is Niels Bohr's son.

Nazi extermination camps.

Heisenberg had left the German nuclear research establishment at Hechingen two weeks before the American team under a Colonel Boris Pash arrived to seize them. At Hechingen, Pash captured several prominent scientists including Otto Hahn, Carl von Weizsäcker, and Max von Laue. From these he learned that Heisenberg was with his family, either in Munich or at a cottage in Urfeld on the Walchensee in the Bavarian foothills.

Heisenberg had bicycled from Hechingen to Urfeld in three days. To avoid strafing by planes he traveled mostly at night and slept in haystacks during the day. When he reached his family he found them safe. For the next week they dragged and piled sandbags around the basement windows and scrounged for food. Most of their neighbors had fled to the other side of the lake. When Colonel Pash finally arrived at Urfeld to take him into custody, Heisenberg said that he "felt like an exhausted swimmer setting foot on firm land."

The captured physicists were interrogated and shipped to England to Farm Hall, a country estate. Here the rooms had been previously wired and all conversations were recorded. There is a somewhat amusing recording of an interchange between Kurt Diebner and Werner Heisenberg.

Diebner: "I wonder whether there are microphones installed here?"

Heisenberg: "Microphones installed? (Laughter) Oh, no, they're not as cute as all that. I don't think they know the real Gestapo methods; they're a bit old-fashioned in that respect."

In 1946, Heisenberg was released and returned to Germany. Here he began the laborious work of reconstruction. With his colleagues he reorganized the Institute for Physics in Göttingen. In an address *Science as a Means for Understanding Among Nations* to the students of Göttingen he made the following statement. "If science is to contribute to understanding among nations, then it is not through its practical importance, neither through the benefits it may provide for someone sick, nor through the terror with which science wrests recognition for a political power but, rather solely from directing attention to that central domain from which the world as a whole obtains order, perhaps simply from the fact that science is beautiful."

In 1948 the Institute for Physics in Göttingen was renamed the *Max Planck Institute for Physics* and in 1955 Heisenberg's wish to live in Munich came to pass. The institute was moved to Munich and was again renamed the *Max Planck Institute for Physics and Astrophysics*.

In 1953, Heisenberg became the president of the *Alexander von Humboldt Foundation*, whose purpose is to bring promising young foreign scholars to Germany. Heinrich Pfeiffer was secretary general of the foundation and became a close friend of Heisenberg. At the Bamberg meeting to celebrate the 100th anniversary of the birth of Heisenberg, Dr. Pfeiffer told several stories about Heisenberg. According to him, two Heisenberg quotations are, "The best government is one you don't notice." Also, after some ill-considered spending cuts were suggested by the government Heisenberg stated, "Only a stupid farmer eats his best potatoes after a bad harvest." According to Helmut Rechenberg, music was second only to physics at the

Symposium On the Development of THE PHYSICISTS' CONCEPTION
OF NATURE at the International Centre for Theoretical Physics,
Miramare, Trieste, Italy — Dedicated to P.A.M. Dirac on his
seventieth birthday.
Werner Heisenberg reminiscing on, "Developments of concepts in the history of
quantum mechanics".

Heisenberg home. In fact, Heisenberg was first attracted to his wife by her music. At Christmas the children presented concerts. Heisenberg was a sufficiently skilled pianist to perform with professionals. Even when he made mistakes the music was always good. In Göttingen there were frequent chamber music concerts at which he played the piano with the two older boys. Chamber music and singing were an integral part of the Heisenberg family. Otto Hahn knew this and when he visited the Heisenberg home he signed the guest book, "Otto Hahn, Tenor."

At the same Bamberg meeting Heisenberg's son, Martin, told several stories about his father. The first two cars that their family owned were named "Archibald" after an old song and "Hannibal" because it was always overloaded for trips over the Alps.

In playing games, their father was very competitive, but tried to keep the games fair for his children. He made it a rule to spend Sundays with the family and took the children outdoors. If he won at a game in which a large element of luck was involved he would say, *"Wie sich Glück und Geschick verketten wird dem Toren niemals klar."* (To a fool it will never be clear how luck and skill interlink.)

He loved to tell jokes and did so frequently, but never at the expense of others. Also, he always tried to put himself at the same level as those around him. In Bavaria he would switch to the local dialect, Bavarian, from High German.

As already mentioned several times, Heisenberg was ever a gentleman. The following episode may illustrate this. When he was invited to Athens to give a semi-popular talk on physics, Queen Frederika of Greece was in the audience. She, like Heisenberg, was originally Bavarian. When all the physicists filed in, only Heisenberg bowed to the queen.

Heisenberg's uncanny ability to perform difficult calculations is legendary. With the advent of electronic computers, Heinrich Mitter (1929 –) wrote a lengthy computer program to do a four-fold integration required in a solution of Heisenberg's theory. The result disappointed Heisenberg since it did not yield what he had expected. He found this hard to believe and asked Mitter if there might not be an error in the computer program. Mitter assured him that he had checked the program very carefully. The next morning Heisenberg said, "There is a mistake. I have done the calculation and the result is as it should be." Mitter and others could not believe that anyone could have done such a complicated calculation without a computer, let alone overnight, and asked Heisenberg to show them how he did it. He then showed them an approximate numerical integration technique that he had learned years before. Now, Mitter rechecked his program and found a mistake. After the program was corrected, the computer calculation agreed with Heisenberg's approximate calculation to within 10%.

For someone who could calculate with incredible accuracy it is surprising that Heisenberg's papers frequently contain mistakes that mysteriously compensate each other. The reason is that he would first do a calculation very carefully one way and, then in writing up the result, would do it in another way so that people might be more easily convinced.

Remember, that although I told Schrödinger's story first, it actually happened almost a year after Heisenberg's discovery. Schrödinger's wave mechanics was greatly welcomed by the "older" physicists, including Wien and Planck, because Heisenberg's formulation had emphasized the discontinuous, quantal nature of quantum mechanics and many hoped that Schrödinger's wave mechanics would provide a return to the more understandable, classical continuous world. Physicists to a large extent still believed Leibnitz' motto, *"Natura non fecit saltum"* (nature does not make jumps). As George Uhlenbeck later stated, "The Schrödinger equation came as a great relief, now we no longer had to learn the strange mathematics of matrices." Einstein also wrote to Paul Ehrenfest about Heisenberg's paper, "In Göttingen they believe in it. I don't." Even Bohr noted that it is "a step probably of fundamental importance" but "it has not yet been possible to apply the theory to questions of atomic structure."

However, after Pauli used the new Matrix Mechanics and found the solution for the spectrum of the hydrogen atom in a crossed electric and magnetic field all that changed. It remained only to distill the essence of Heisenberg's ideas.

Chapter 12

Matrix Mechanics is Born

"Physics is too difficult for physicists", David Hilbert.

Max Born (1882 – 1970), Pascual Jordan (1902 – 1980) and Werner Heisenberg soon distilled the essence of matrix mechanics. Born's birthplace was then the Prussian (Silesian) city of Breslau (now Wroclaw). While he was still a small boy, his mother died and so his main influences, while growing up, were his father and his classmates. His father had done research in embryology and the mechanics of evolution. His classmates, Otto Töplitz and Ernst Hellinger, later became famous mathematicians.

After finishing the *Gymnasium*[1] he went to Göttingen where the "three prophets": David Hilbert (1862 – 1943), Felix Christian Klein (1849 – 1925), and Hermann Minkowski (1864 – 1909) preached. Born was a very good student. However, his relations with the mathematician, Felix Klein, were not good; Klein's lectures were too polished. So, Born skipped them and this led to several problems. Due to a classmate's illness Born learned only with short notice that he was due to give a report on a problem in elasticity. Since he no longer had sufficient time to study the literature, he developed his own ideas. This impressed Klein and he suggested this problem for the annual University Prize and wrote to Born that he expected him to submit a paper. Although reluctant, Born submitted a paper and won. Thereafter Born switched from mathematics to astronomy in order not to have to be examined by Klein in whose bad graces he remained.

On his final exam in 1906, the astronomer, Karl Schwarzschild (1873 – 1916) asked Born, "What would you do if you saw a falling star?" Born

---

[1]The *Gymnasium* is roughly the equivalent of high school, although students usually enter the Gymnasium after grade four or five.

replied confidently, "I'd make a wish." Hilbert laughed, but Schwarzschild repeated, "So you would, would you? But, what else?" By now Born had collected his thoughts and explained in great detail all the observations he would make. Incidentally, this is the same Schwarzschild after whom the "radius" of a black hole is called the Schwarschild radius.

It was then quite common for students to switch schools several times in order to hear from different professors. Thus, in 1907 Born went as an advanced student to Cambridge, to take a course in electricity. A pretty young lady from Newsham, who seemed shy and standoffish to Born, but who himself was actually the shy one, was taking the same course with him. The instructor in this course was an old professor with an impish sense of humor. One day when Born and his lab partner were having some difficulty with their equipment, Born asked the professor for some help, "Dr. Searle, something is wrong here. What shall I do with this angel?" Of course he meant "angle". Old Searle peered at both of them over his spectacles, wagged his head and said, "Kiss her, man kiss her." After that Born's shyness was even greater.

As a student Born lived in "House El Bokarebo" which was named after the first two letters of its inhabitants' names: Ella Philipson, Born, Theodore von Kàrmàn, Albrecht Renner, and Hans Bolza. Born had received a telegram from von Kàrmàn, "Will you share a house with me and three other lunatics?" Born and von Kàrmàn had often planned to rent a comfortable house with a well-trained housekeeper. When von Kàrmàn finally found the ideal housekeeper, there was a problem. Sister Annie was an ex-nurse and wanted to open a house for the elderly and infirm. Von Kàrmàn finally persuaded her to accept by convincing her that although neither elderly, nor infirm they were all somewhat insane.

Theodore von Kàrmàn (1881 – 1963), was an engineer by training, but did very original work in fluid dynamics. When other engineers told him that they could not follow his fancy mathematics since they were only "practical engineers" his reply invariably was, "Yes, you know what a practical engineer is? He is someone who perpetuates the mistakes of his predecessors." He also stated that, "The scientist describes what is; the engineer creates what never was."

Since he was slightly deaf, von Kàrmàn needed a microphone and earphones to hear. However, he used this handicap to advantage, for when a speaker was particularly dull, von Kàrmàn had the habit of turning off his microphone and contemplating in peace.

Von Kàrmàn, like Wigner, Teller, and von Neuman, was of Hungarian descent and explained the extraordinary abilities of his countrymen as follows. If a Hungarian follows a normal person through a revolving door, you can expect the Hungarian to emerge first.

The same von Kàrmàn taught Born the essentials of mathematical physics which Born then adopted as his own. They are:

1) Regard the problem in the right perspective.

2) Estimate by rough and ready methods the order of magnitude of the results expected before going into detailed calculations.

3) Use approximations adapted to the accuracy needed, even if no rigorous proof of their validity can be given.

4) Be constantly aware of all the facts.

In 1912 when Born visited Michelson, the latter showed him how to take photographs of spectra. Born was very much impressed by the regularities of bands in the spectrum of carbon. Michelson asked, "Do you really believe that there is a simple law behind it if horrible things like that happen?" Unknown to Michelson, Deslandre's formula already explained these regularities.

While Born was serving with the German military during World War One, he was sent with other scientists and officers to the Baltic for an inspection. On a cold and foggy November they waited on the docks of the small port city of Neustadt until dark. Finally they gave up and repaired to a nice-looking inn. Here some naval officers joined them and they settled down at a big table to do some serious drinking. After a while nature called and Born exited through a small door expecting a toilet. The door snapped shut behind him and he found himself trapped; the door was gone. So he sat down to think how it could be that he was walled in. After what seemed an eternity light poured in. His friends had missed him and come looking. They found him sitting in the space inside a spiral staircase.

About 2:00 AM, the destroyer came and they embarked. It was a cold, rough journey so they had to sit in the officer's mess. This led to an attack of asthma. Later they transferred to a big liner used as an auxiliary cruiser and experimental station. There, the drinking continued and the asthma got worse. A visit to a submarine aggravated Born's situation. The captain seeing Born's distress offered his chart room on the bridge, "for some fresh air". The cold drove Born back inside and eventually he wound up in the ship's infirmary. Since by this time they were near Warnemünde, Born and his colleagues were taken ashore and so ended his military expedition.

Also during World War One, the German ambassador in Stockholm sent two Swedish inventors to Berlin. These two gentlemen were lodged in the Adler, the best hotel in Berlin, and all their expenses were covered by the German government. Clearly they were not too anxious to leave. Ostensibly they had invented an ammunition detector. It was the job of the scientists in the military, namely Born, to evaluate their discovery.

At first these two gentlemen claimed that they needed ten kilograms of platinum. After much discussion this was reduced to a more reasonable amount. Next, they refused to have their instrument tested anywhere except at the front where, as Born repeatedly pointed out, no controlled experiment was possible. Thus, weeks went by with these two enjoying the hospitality of the Adler. Finally they agreed to tests at an experimental station.

A hundred boxes, of which two contained ammunition and ninety-eight contained sand, were arranged in a circle. Many high-ranking officers were present to witness the great event. After a long mysterious preparation the men began to turn their instrument with an arrow on it and, sure enough, found one of the boxes of ammo. The general shouted, "Well done!" and turning to the scientists, among which was Born, "Your skepticism is for once disproved. I think the test is over."

It took much persuasion by Born to convince the general to repeat the test once more. This time the test failed as it did in all subsequent cases. In the end to show that no swindle had occurred, Born had to open all one hundred boxes to show that two contained ammunition. The general had by now left for more urgent business.

Early on in his career Born gave up experimental physics. At one point he had attempted to do an experiment on black-body radiation using a ceramic oven surrounded by water cooling. Unfortunately he forgot to turn off the water when he left and he not only flooded his lab but also that below his. Otto Richard Lummer (1860 – 1925), the occupant of the laboratory below was not amused, but Ernst Pringsheim (1859 – 1917) whose lab was also flooded composed a poem and Born managed to escape by paying for the damages. Lummer and Pringsheim had also collaborated on the very precise measurements of blackbody radiation discussed earlier.

What really convinced Born to give up experimental work was Robert Pohl's (1884 – 1976) obsessive fear of mercury poisoning. After getting to Göttingen, Born decided to do some experimental work together with a student. This time the laboratory was above Pohl's study and one day a fine drizzle of mercury rained from the ceiling onto the floor of Pohl's study. A big vessel of mercury from a vacuum pump had broken. Pohl was livid with anger and fear. He ordered the floor of his poisoned office to be removed. At this point Born decided to give up experimental work for good.

When Born gave a lecture at the Cavendish, Rutherford made the following statement to him. "There must be an experimental physicist with exactly the same name as you. Because, when I was preparing a lecture on the kinetic theory of gases I found a paper in the *Physikalische Zeitschrift* signed by Max Born and Elisabeth Bormann and it contained the description of an experiment much too good to have been written by a mathematician like you." Actually there was only one Max Born and the experiment referred to dealt with the measurement of molecular cross-sections by measuring the intensity of a beam of silver atoms in vacuum and in gas.

Like all prominent scientists, Born was called upon to referee more than his share of papers submitted for publication in scientific journals. After rejecting a paper that had arrived at an obviously incorrect result, the author resubmitted the paper with the query where the error was. An exasperated Born was heard to remark, "I am not here to find somebody else's mistakes."

After the Nazis came to power Max Born, who was Jewish, left his native Germany and accepted an invitation to Cambridge. His wife Hedwig used to quote an old Jewish saying, "An anti-Semite is someone who hates Jews more than absolutely necessary."

It may also have been around this time that Born said, "The belief that there is only one truth, and that oneself is in possession of it, is the root of all evil in the world."

And again, later in *My Life and My Views* he wrote, "It seems to me that the attempt made by nature on this earth to produce a thinking animal may have failed." In the same book he stated, "There is no longer any doubt: all matter is unstable. If this were not true, the stars would not shine, there would be no heat and light from the sun, no life on earth. Stability and life are incompatible. Thus life is necessarily a dangerous adventure which may have a happy ending or a bad one."

On his arrival in the Cambridge railway station Born was severely shocked to see a gigantic poster proclaiming

BORN TO BE HANGED

The people from Cambridge calmed him, however, by explaining that this was only an advertisement for a play, "The man born to be hanged".

To get back to matrix mechanics. On the North-Express from Göttingen to Hanover, to attend a meeting of the Lower Saxony branch of the German

Physical Society, Born met several physicists including Wolfgang Pauli, who had been his assistant in 1922. Born told him about the fantastic break-through that Heisenberg had made in quantum physics and that he had succeeded in evaluating in quite general form the diagonal part of some of Heisenberg's tables or matrices. However, he complained about the extraordinary difficulty of finding all the off-diagonal elements and asked Pauli if he would like to collaborate with him on this problem. Instead of the expected interest he got a cold sarcastic refusal, "Yes, I know you are fond of tedious and complicated formalisms. You are only going to spoil Heisenberg's physical ideas by your futile mathematics."

For his biting criticisms Pauli came later to be known as, "God's self-appointed vicar on earth". Regarding Born's approach, Pauli wrote to Kronig, "One must attempt to free Heisenberg's mechanics from the Göttingen *Gelehrsamkeitsschwal* (tidal wave of erudition)."

Pauli had complained to Heisenberg in a similar vein about the excessively formal approach of the Göttingen school. To this Heisenberg responded in a somewhat strong tone. What follows is my somewhat free translation. "With regard to your last two letters I need to preach to you and excuse me if I have to continue in Bavarian: It is truly a pigsty that you are unable to stop the rabble rousing. Your continuous abuse of Copenhagen and Göttingen are a screaming scandal. You'll have to concede to us that, in any case we are not trying to destroy physics with bad intentions; if you point out to us that we are such big asses that we might never achieve anything physically new, you may be right. But in that case you are just as big an ass, because you also don't achieve it ... (The dots represent swearing for about two minutes!) Don't take it personally and many greetings."

A young man, Pascual Jordan (1902 – 1980) sitting in the same compartment had overheard the conversation between Born and Pauli and at the station in Hanover introduced himself to Born. He told Born of his experience in working with matrices and that since he was presently unemployed, expressed his willingness to work as Born's assistant. Jordan, then found the solution. All the off-diagonal elements, that Born had not been able to compute, were zero. This led to a most fruitful collaboration culminating in the famous Born-Jordan paper. In this paper the essence of Heisenberg's idea was distilled. Born and Jordan found that the essential content of the matrices was such that the difference between the product of the matrices for position times momentum and momentum times position was just Planck's constant h (actually $ih/2\pi$), a truly beautiful result. This equation reads simply

$$qp - pq = ih/2\pi. \tag{12.0.1}$$

It contains all the quantum conditions. Born and Jordan furthermore went on to show that energy conservation, which Heisenberg had so laboriously verified on Heligoland for a simple oscillator, was also true for a large variety of systems.

A month later (November) Heisenberg, who also was not entirely happy with the Born-Jordan approach, wrote to Pauli. "I'm still pretty unhappy about the whole theory, and was thus glad to see that you were so completely on my side in your views on mathematics and physics. Here I'm in an environment that feels and thinks the exact opposite, and I do not know if I'm just too stupid to understand mathematics. ... Göttingen splits into two camps, one which with Hilbert ... speaks of great success, achieved through the introduction of matrix rules to physics, the other which with Franck says that we still cannot understand the matrices."

The influence that Born had on Jordan was profound, as Jordan himself stated in a eulogy to Born. "He was not only my teacher, who in my student days introduced me to the wide world of physics—his lectures were a wonderful combination of intellectual clarity and horizon widening overview. But he was also, I want to assert, the person who next to my parents, exerted the deepest, the longest lasting influence on my life."

Actually, Born's influence extended to many famous students among which J. Robert Oppenheimer and Arthur H. Compton rank as two of the foremost. Later Compton wrote, "In the winter of 1926 I found more than twenty Americans in Göttingen at this fount of quantum wisdom."

After this labor, Born took a well-earned vacation in Switzerland. On his return, he recommenced work with Heisenberg and Jordan. This led to the so-called "Three-men paper" in which all the fundamental principles of modern matrix mechanics were expounded. As Heisenberg later recalled, "It was the particular spirit of Göttingen, Born's faith that nothing short of a new self-consistent quantum mechanics was acceptable as the goal on fundamental research that enabled (my) ideas to come to full fruition."

Shortly after the publication of the famous Three-men paper by Born, Heisenberg and Jordan on matrix mechanics, Heisenberg found himself traveling together with Harald August Bohr (1887 – 1951) and Godfrey Harold Hardy (1877 – 1947) on the train from Bohr's cottage at Tisvilde to Copenhagen. Niels Bohr's brother Harold was a very prominent mathematician, rivaling his brother in fame. Thus, it happened that mathematicians were also frequent guests in the home of Niels Bohr. Hardy was reputed to be one of the purest of pure mathematicians. He once told Bertrand Russell "If I could prove by logic that you would die in five minutes, I should be sorry you were going to die, but my sorrow would be very much mitigated by pleasure in the proof." He also stated that, "It is not worth an intelligent man's time to be in the majority. By definition, there are already enough people to do that." Also, after working with abstract group theory, he is supposed to have declared, "Finally, a piece of pure mathematics that will not be sullied by applications." Within two years he was proven wrong, so much so, that in his 1928 book *Gruppentheorie und Quantenmechanic* (Group Theory and Quantum Mechanics) Hermann Weyl (1885 – 1955) was able to write in the preface, "It has been rumored that the 'group pest' is gradually being cut out of physics."

In the early thirties Hardy mused about the lack of respect for mathematics and science. "It's rather odd but when we hear about intellectuals nowadays, it doesn't include people like me and J. J. Thomson and Rutherford."

Hardy with a touch of sarcasm, presented Heisenberg with, "a little mathematical problem for practice." It was the theory of a Chinese game and had just recently been solved. While Heisenberg struggled with it Harold Bohr chided Hardy for "using a young man's mathematical force on such nonsense." At this point Heisenberg found part of the theory and presented it to Hardy. The latter, ever skeptical, replied dryly, "Well, your new atomic theory may at least work for the hydrogen atom."

Regarding group theory, James Jeans had also expressed his opinion early in the 1900's to the mathematician Oswald Veblen. After he had looked over the Princeton curriculum he declared, "Group theory will never be useful."

Pauli, in an incredibly arduous computation, used matrix mechanics to derive the hydrogen spectrum and thus showed that matrix mechanics was at least as good as the Old Quantum Theory. He also showed that matrix mechanics was, in fact, better than the Old Quantum Theory in solving the problem of a hydrogen atom in a crossed electric and magnetic field. By skillfully wielding the tools of matrix mechanics, Pauli vanquished this problem, which had defied all efforts with the Old Quantum Theory. That a correct quantum mechanics had been found could now no longer be doubted. What was still needed was a consistent physical interpretation.

Born and Jordan seem to have carried their predilection for matrix mechanics too far. Most problems are solved more simply by using the

Schrödinger equation. Nevertheless, in 1930 after quantum mechanics as well as the Schrödinger equation were well-established, they published a book in which they used nothing but matrix mechanics. Pauli, in his usual straightforward manner, wrote a rather sarcastic review. "Many of the results of quantum theory can in fact in no way be deduced by means of these so-called elementary procedures, others only inconveniently and with indirect methods. (To these latter ones belongs for instance the determination of the Balmer terms, which is carried out by matrix calculus, along the lines of an earlier paper by Pauli on this subject. So one cannot reproach the reviewer with finding grapes sour because they are too high.)" The final sentence of Pauli's review especially displeased Born, "... print and paper are excellent."

Later in life, Born confessed, "I never liked being a specialist and have always remained a dilettante, even in what were considered my own subjects."

In 1954 during a visit to Princeton, Pauli told Freeman Dyson, "If I had not wasted so much time trying to make sense of five-dimensional relativity (Kaluza-Klein theory and similar attempts) I might have discovered quantum mechanics myself."

At a tea in 1929 Edward Teller sat next to Bohr and speculated that, "Someday matrix mechanics will be the common way of thinking. This will indeed be a benefit because then the contradictions will disappear from the orderly discussions of science." During this speech Bohr's eyelids closed and after a long pause he whispered with his eyes still closed, "You might as well say we are not sitting here at all but merely dreaming it."

Heisenberg received the Nobel Prize for this work. Although pleased to have received the prize, Heisenberg was unhappy that Born had not been recognized and sent him a letter in which he expressed these sentiments. Born's work was belatedly recognized by the Nobel Committee in 1954 when he was already in his seventies. Born's citation read, "for his fundamental research in quantum mechanics, especially for his statistical interpretation of the wavefunction." He shared the prize with Walther Bothe (1891 – 1957) whose citation read, "for the coincidence method and his discoveries made therewith".

The mathematician, David Hilbert, also played an interesting part in all this. In the summer of 1925 shortly after developing matrix mechanics and, before the Schrödinger equation had been found, Max Born and Werner Heisenberg went to see Hilbert to get some help with their matrix mechanics. After listening to them Hilbert remarked that in his experience these large matrices seemed to be always something that resulted from calculating with a wave equation of some sort. He advised them to look for the associated wave equation. This would greatly simplify their work. According to Edward Condon they thought that this was a silly idea and that Hilbert was either pulling their legs or did not know what he was talking about. Apparently Hilbert had great fun teasing them after Schrödinger published the equation named after him.

Hilbert was also rather hard on mathematicians for "lacking imagination". He felt that mathematicians should have discovered relativity, but instead it was that physicist, Einstein. Similarly, he stated that he had worked hard at understanding atomic spectra and again it was some physicists that had come up with the answer.

Once a year at Göttingen, Felix Klein would give a seminar followed by a banquet at some major industrial establishment. One year, after the date was already set, he fell ill and asked David Hilbert to replace him. He repeatedly impressed upon Hilbert that it was necessary to emphasize the close relationship between technology and mathematics. The speech that Hilbert then delivered carried essentially the following message. Mathematics and technology are on the best possible terms with each other, not only now, but also for the foreseeable future since they have nothing to do with each other.

At a party attended by Hilbert, talk about astrology arose. Some of the people argued that even if science could not justify the influence of the stars on human affairs neither could science disprove it. Hilbert listened for a while but kept repeating his characteristic, *"Na, ja. Na, ja."* meaning, "Oh, well. Oh, well." so often that everyone fell silent and listened. At this point he said in his most sarcastic and piercing voice, "If you were to appoint a committee of the twenty cleverest people in the world, I say the acknowledged supreme brains of every nation and give them the task of finding the most stupid thing possible, they would fail to discover astrology. *Na, ja. Na, ja.*" That ended the conversation about astrology.

Hilbert had a criterion for the quality of research. "One can measure the importance of a scientific work by the number of earlier publications rendered superfluous by it."

Also, when someone blamed Galileo for not standing up for his convictions Hilbert became quite irate and said, "But he was not an idiot. Only an idiot could believe that scientific truth needs martyrdom; that may be necessary in religion, but scientific results prove themselves in due time."

On Hilbert's tombstone in Göttingen two of his sentences are inscribed:

Wir müssen wissen.   (We must know.)
Wir werden wissen.   (We shall know.)

When the new Nazi Minister of Education asked David Hilbert if the Physics Institute in Göttingen had suffered from "the departure of the Jews and their friends" he replied, "Suffered? No, Herr Minister, it didn't suffer; it just doesn't exist any more."

Similar sentiments were voiced at other German universities. At the University of Hamburg many of the top physicists and mathematicians had also either left or been forced to leave, due to their Jewish heritage or sympathies. The president of the University was a staunch Nazi from the faculty of medicine. During a university reception W. Lenz stood up and addressed the president, "Herr Rektor, the fact that there are so many on the medical faculty and so few physicists is not a coincidence." With these words he sat down again.

Although quantum mechanics had been started and completed in Germany, the Nazis had ensured that German science would now be set back for many decades.

Chapter 13

The Purest Soul's Beautiful Quantum Mechanics

"It has become increasingly evident in recent times, however, that nature works on a different plan. Her fundamental laws do not govern the world as it appears in our mental picture in a very direct way, but instead they control a substratum of which we cannot form a mental picture without introducing irrelevancies." P.A.M. Dirac, Preface to the First Edition of Quantum Mechanics.

Shortly after Heisenberg's return from Heligoland where he had come up with the first results in the direction of matrix mechanics, he was invited to present a lecture to the Kapitza Club at the Cavendish Laboratory in Cambridge. The title of his talk was *Term Zoology and Zeeman Botany*. At the end of his talk he briefly mentioned his new results on matrix mechanics. In the audience was another young physicist, Paul Adrien Maurice Dirac (1902 – 1984) who, although this work should have been of extreme interest to him, paid no attention to it. As Dirac himself later explained, the seminars at the Cavendish Laboratory were held at two o'clock in the afternoon. "After a good lunch, this time was more suitable for a snooze than concentrated attention on a speaker."

About six weeks later Dirac's colleague, R. H. Fowler (1889 – 1944), obtained the proofs of Heisenberg's paper from Bohr and passed them on to Dirac. This time Heisenberg's ideas fell on fertile ground. Guided partly by the success of the old Bohr-Sommerfeld theory and largely by his own esthetic sense, Dirac concentrated on Hamilton's formulation (especially something called Poisson brackets) of classical mechanics.

According to Dirac, he first hit upon the idea of using classical Poisson brackets on a weekend, but could not remember the exact definition of these objects and had to wait until Monday for the library to open so that he could verify his ideas. Within weeks, he succeeded in demonstrating a very deep relationship between Poison brackets and what are now called commutation relations, as well as Hamilton's formulation and Heisenberg's matrix mechanics. Thus, only a few months after Heisenberg's paper, there appeared Dirac's first of a series of papers in December 1925. Schrödinger's four papers, including the proof that his and Heisenberg's formulation were equivalent, all appeared somewhat later, with the last one in September 1926.

Symposium On the Development of THE PHYSICISTS' CONCEPTION OF NATURE at the International Centre for Theoretical Physics, Miramare, Trieste, Italy — Dedicated to P.A.M. Dirac on his seventieth birthday.
Paul A.M. Dirac writing some of the fundamental constants of nature.

Dirac proceeded from Heisenberg's results and built, in a series of brilliant papers, an alternate mathematical framework that was far more appealing to physicists. It was neither wave nor matrix mechanics, but a universal language which encompassed both. It was Quantum Mechanics.

These papers were so important, in giving physicists a practical computational tool, that when in 1930 Dirac published a book on the subject, another prominent physicist, Sir John Edward Lennard-Jones (1894 – 1954) made the following remark. "An eminent European physicist who is fortunate enough to possess a bound set of reprints of Dr. Dirac's original papers, has been heard to refer to them affectionately as his 'bible'. Those not so fortunate have now, at any rate, an opportunity of acquiring a copy of the authorized version."

According to Hans Bethe, Dirac, when lecturing, read from his book in his course on quantum mechanics. When asked why he did this he replied, "When I wrote the book I thought very hard about how to formulate it, and there is no better way." This seems to agree with Hendrik Casimir's recollections of how Dirac answered questions when, in 1928 while completing his book, he lectured at Leiden. "The lectures were beautifully clear, and when Ehrenfest or somebody else asked for a further explanation, Dirac would just repeat what he had said." In a similar manner when Ehrenfest and his assistant A. J. Rutgers wrote to Dirac for an explanation of one of his papers the response was almost identical to what was originally written in the paper. After further study of the paper, Ehrenfest commented, "The better one understands it, the better it is put down there." So, maybe Dirac was right and there was no better way to say it.

Paul Dirac's father, Charles Adrien Ladislas Dirac, was a Swiss citizen who came to England in about 1888 and taught French at Bristol. In 1899 he married Florence Hannah Holton. They had three children: Reginald Charles Felix, Paul Adrien Maurice, and Beatrice Isabella Marguerite Walla. The father was extremely strict and managed to alienate all three of his children. This had lasting effects on Paul.

Paul Dirac was also always very precise in his speech. In fact, when he presented a talk it could be printed verbatim. He explained his precise speech as follows. "My father made the rule that I should only talk to him in French. He thought that it would be good for me to learn French that way. Since I found that I could not express myself in French, it was better for me to stay silent than to talk in English. So I became very silent at that time and that started very early." This quotation helps to explain both Dirac's precision and taciturnity in his speech as well as his dislike for his father. When Dirac was informed of the death of his father he did not return for the funeral, but after the funeral wrote to his wife, "I feel much freer now." By the way, his father's technique worked and Dirac learned to speak flawless French.

Throughout his life Dirac avoided going to Switzerland, a country that he associated with his father. Dirac's mother and his older brother Reginald and his sister never managed to live up to the father's standard. So they always ate apart in the kitchen. Also there was never any social occasion at their house. In Dirac's words, "I had no social life at all as a child. No one ever came to our house for social purposes."

At age twelve Paul Dirac entered the secondary school where his father taught. World War One started and the older boys left for military service. This meant that the younger boys had better access to all facilities, especially the science laboratories. He wrote of these days at the Merchant Venturers Technical College. "The Merchant Venturers was an excellent school for science and modern languages. There was no Latin or Greek, something of which I was rather glad, because I did not appreciate the value of old cultures. I consider myself very lucky in having been able to attend the school. ... I was rushed through the lower forms, and was introduced at an especially early age to the basis of mathematics, physics, and chemistry in the higher forms. In mathematics I was studying from books which mostly were ahead of the rest of the class. This rapid advancement was a great help to me in my latter career."

In 1918 he entered Bristol University to study electrical engineering. After finishing he worked during the summer of 1923 as an engineer, but this did not appeal to him and he decided to continue with mathematics at Cambridge. Although he won a scholarship to St. John's College Cambridge, he could not attend since the scholarship did not provide enough funds. The local education authority was expected to supply the extra funds, but declined because Paul's father had not been a British citizen long enough. Paul therefore studied mathematics for one year at Bristol. The following year he was awarded a research grant at Cambridge and started his research under Ralph Fowler (1889 – 1944).

Later in life, Dirac commented about his education as an engineer. "I owe a lot to my engineering training because it [taught] me to tolerate approximations. Previously to that I thought ... one should just concentrate on exact equations all the time. Then I got the idea that in the actual world all our equations are only approximate. We must just tend to greater and greater accuracy. In spite of the equations being approximate, they can be beautiful."

His brilliance was immediately obvious. Within six months of entering Cambridge he published two papers on statistical mechanics. These were followed in 1924 by his first paper on a quantum problem and four more papers in 1925.

However, tragedy struck. His brother Reginald committed suicide and Paul seems to have ascribed a large measure of blame to his father. His relations with his father, already strained, became even more so and Paul became more withdrawn.

When Dirac received the Nobel Prize, his mother was present, but his father was not invited.

The alienation between father and son can also be seen from the following. In the early 1930s, Professor Tyndall gave a series of public lectures on modern physics. A considerably older man attended regularly and sat in the front row taking careful notes. After the last lecture this man came up to Tyndall to thank him and said, "I am glad to have heard all this. My son is also a physicist, but he never tells me anything about what he does." The man was Charles Dirac, father of Paul Dirac.

In constructing his general form of quantum mechanics Dirac invented some totally new mathematics. Of course mathematicians did not take this lightly and Johann (John) von Neumann (1903 – 1957) in his book *Mathematical Foundations of Quantum Mechanics* states in the introduction, "The method of Dirac ... in no way satisfies the requirements of mathematical rigor." Again a little further on in reference to the Dirac delta function, "The insertion of such a mathematical 'fiction' is frequently necessary in Dirac's approach."

Since Dirac's version did not conform to the mathematics of the time, von Neumann's attack in his book *Mathematical Foundations of Quantum Mechanics* that the so-called Dirac delta function was based on a "mathematical fiction" was justified for the moment. Von Neumann was correct, no such function exists. Nevertheless, Dirac's formulation had such practical computational advantages that physicists continued to use it and to obtain results in concordance with those obtained by the mathematically acceptable formulation of von Neumann. Finally in 1945, the mathematician Laurent Schwartz (1915 – 2002) settled the impasse and vindicated Dirac's approach by showing that, although the Dirac delta function cannot possibly exist as a function, it can nevertheless exist as a perfectly respectable mathematical entity called a distribution. This opened a whole new branch of mathematics, the theory of distributions. So much for the mathematical niceties. As one of my fellow graduate students, J Peter Vajk, once told me when I criticized his non-rigorous approach, "Rigor impedes vigor."

Von Neumann's reproach is understandable since he was the consummate mathematician. When a student asked him for the prerequisites for his course on *The Foundations of Quantum Mechanics* he replied, "All of mathematics." Regarding mathematics, he had this to say. "In mathematics you don't understand things. You just get used to them." Another statement that he uttered on more than one occasion was, "A good formalism is one that you can apply without thinking." This is not so different from a statement by Mambillikalathil Govind Kumar Menon (1928 –) in 1967. "A first-rate laboratory is one in which mediocre scientists can produce outstanding work."

Later von Neumann was one of the leaders in developing electronic computers. By 1949 these had advanced as much as vacuum tube technology made possible. This led him to say, "It would appear that we have reached the limits of what it is possible to achieve with computer technology, although one should be careful with such statements, as they tend to sound pretty silly in five years."

Regarding the use of such computers to generate random numbers he pronounced, "Anyone who attempts to generate random numbers by deterministic means is, of course, living in a state of sin."

Von Neumann belonged to a group of Hungarian physicists, all of whom graduated from the same high school in Buda, one half of the city of Budapest. There are at least two proponents of the theory that these physicists were really settlers from Mars. To support this theory they point to the language that these people speak, the incredible beauty of their women (the Gabor sisters), their superhuman intelligence, and their music (Béla Bartók). The two proponents of the theory are: Fritz Houtermans and Philip Morrison. The Martians are: Theodore von Kármán, John von Neuman, Edward Teller, Leo Szilard, and Eugene Wigner. When William Sears introduced von Kármán to Morrison, who presented him with this theory, von Kármán chuckled, "Funniest thing I have ever heard. Mind you, I don't deny it."

Although the mathematical abilities of von Neumann are well known, his ability to consume alcohol is less well known. On one occasion, on a bet, he downed sixteen martinis in a row and remained on his feet, to all appearances quite lucid, although more pessimistic than usual.

According to Edward Condon (1902 – 1974), at a physics lecture in Princeton an experimentalist was presenting data with many points badly scattered. The speaker then attempted to convince the audience that these points lay on a curve. John von Neumann was heard to mutter, "At least they lie in a plane."

Von Neumann once rushed up to his friend, Alice Eppinger, and asked her for five Pfennig for a streetcar. She replied that she did not have five Pfennig, but offered him a Mark. He looked at the proffered coin for quite some time, frowned and shook his head. "No, no, I only need five Pfennig." and walked away.

Also in his youth he was asked at what age of a mathematician would cease to be productive. He answered that it would be around 50. When he was fifty he was again asked. This time he revised the age upwards by quite a bit.

Something else that made Dirac's way of doing quantum mechanics so appealing to physicists was his notation. He had found a very elegant and compact way of writing the formulas of quantum mechanics. With his whimsical sense of humor he had called the fundamental quantities $\langle|$ and $|\rangle$ "bra" and "ket" because when they are put together to give a number they form $\langle | \rangle$, a "bracket".

Dirac, throughout his life, was guided by his own sense of esthetics. Thus, it is not surprising to have the following quotes from him. "A theory that has some mathematical beauty is more likely to be correct than an ugly one that gives a detailed fit to some experiments." And on another occasion, "The mathematics that applies to physics must be beautiful since it is the form selected by God." It may have been his love of mathematical formalism that led Dirac to warn Schrödinger, "Beware of forming models or mental pictures at all."

The following quote is from "The Evolution of the Physicists' Picture of Nature," *Scientific American 208*, 5, 43-53, (1963). "It seems to be one of the fundamental features of nature that fundamental physical laws are described in terms of a mathematical theory of great beauty and power, needing quite a high standard of mathematics for one to understand it. You may wonder: why is nature constructed along these lines? One can only answer that our present knowledge is that nature is so constructed. We simply have to accept it. One could perhaps describe the situation by saying that God is a mathematician of very high order, and He used very advanced mathematics in constructing the universe. Our feeble attempts at mathematics enable us to understand a bit of the universe, and as we proceed to develop higher and higher mathematics we can hope to understand the universe better."

These statements are a little surprising since in 1927 at the Solvay conference Dirac spoke out strongly for atheism, asserting that God is simply a human invention and religion is still being taught to keep the masses more docile. A fierce discussion followed and Pauli summarized by saying, "Our friend Dirac too has a religion, and its guiding principle is, 'There is no God, and Dirac is his prophet.' "

Symposium On the Development of THE PHYSICISTS' CONCEPTION OF NATURE at the International Centre for Theoretical Physics, Miramare, Trieste, Italy — Dedicated to P.A.M. Dirac on his seventieth birthday.
First row, left to right: C. F. von Weizscker, W. Heisenberg and A. Salam.
Second row, left to right: F. Hund, two unidentified persons, E. Wigner.
Third row, second from left: S. Chandrasekhar.
Fourth row, third from left: A. Wheeler.

Dirac applied his reasoning to all aspects of beauty, so it was not surprising that he developed the theory that, "There must be an optimal distance for perceiving the beauty of women." His argument went as follows. If they are very far away all you see is a blob. If they are too close all you see is the craters in the face and all the tiny imperfections. Thus there must be an optimum distance. At the time he developed this theory, Dirac was still unmarried and when George Gamow asked him what was the closest distance at which he had seen a woman he replied, "About this far." with his hands about two feet apart.

According to Freeman Dyson, when someone asked Dirac what he meant by the beauty of a mathematical theory the response was something like, "If the questioner was a mathematician then he did not need to be told, but were he not a mathematician then nothing would be able to convince him of it."

I heard the following two stories when, in 1995, I visited the Tata Institute for Fundamental Research (TIFR) in Bombay (now Mumbai). In 1948 – 49 Dirac visited TIFR for the first time. During a conversation with Subrahmanyan Chandrasekhar, Mrs. Dirac told him, "Paul is becoming impossible."

"How is that Mrs. Dirac?"

"He only walks up and down the streets of Cambridge with his hands behind his back and looks at the sidewalk all the time."

Dirac interjected, "That's not true. When I see a pair of pretty ankles, I look up."

On the same occasion, after the rather tiring plane trip Mrs. Dirac complained to a friend, "We were packed in like sardines."

Dirac was heard to mutter, "They don't pack sardines like that."

Dirac's frugality with words was legendary. A standard joke was that the only words he ever spoke were "Yes," "No," and "I don't know." On one occasion Chandrasekhar was explaining some ideas to Dirac who kept saying, "yes" but added that "yes" simply meant that Chandrasekhar should continue and did not imply that he agreed with what was being said.

Dirac's predilection for few words was even recorded in the *Wisconsin State Journal* when Dirac visited the University of Wisconsin in 1929. The header read,

ROUNDY INTERVIEWS PROFESSOR DIRAC: AN ENJOYABLE TIME WAS HAD BY ALL.

What follows is part of that interview.

The other afternoon I knocks at the door of Dr. Dirac's office in Sterling Hall and a pleasant voice says, "Come in." And I want to say here and now that this sentence "come in" was about the longest one emitted by the doctor during our interview. He is sure all for efficiency in conversation. It suits me. I hate a talkative guy.

"What do you like best about America?" says I.

"Potatoes," says he.

"Same here," says I. "Do you like to read the Sunday comics?"

"Yes," says he warming up a bit more than usual.

"And now I want to ask you something more—do you ever run across a fellow that even you can't understand?"

"Yes," says he.

"This will make great reading for the boys down at the office," says I. "Do you mind releasing to me who he is?"

"Weyl," says he.

Another example of Dirac's taciturnity occurred at a luncheon. Dirac, his brother-in-law, Eugene Wigner, and Michael Polanyi were discussing questions of science, society, and such things. Throughout the meal Dirac did not utter a word. So, after lunch as they went out, Wigner turned to him and said, "Paul, why don't you speak up? Everybody is interested in what you have to say." Dirac responded, "There are always more people willing to speak than willing to listen."

Dirac's precision is further illustrated by the following stories. When in Copenhagen, O. Klein and Y. Nishina gave a blackboard derivation of their formula for electron-photon scattering. The final result differed by a sign from their published result. When this was pointed out to them they agreed that the published result was correct and that they must have made some sign errors on the blackboard. Dirac pointed out, "There must be a sign error in an odd number of places."

With reference to Dostoyevsky's *Crime and Punishment*, one of the few books that Dirac read, he commented, "It is nice, but in one of the chapters the author made a mistake. The sun rises twice on the same day."

Another of the rare books read by Dirac was E. M. Forster's *A Passage to India*. When Dirac met Forster, the story goes that Dirac asked Forster, "What happened in the cave?" Forster's reply supposedly was, "I don't know." However, Rudolph Peierls asked Dirac what really was said. The answer was, "I asked if there was a third party in the cave? The response was, 'No'."

In 1968, I attended the Coral Gables conference where Dirac gave a talk. During the question period someone in the audience stood up and said, "Professor Dirac I don't see how equation such and such follows from equation so and so." After a long silence the questioner who was still standing, asked, "Aren't you going to answer my question?" To this Dirac replied in typical fashion, "That was a statement not a question." Apparently the same sort of thing happened on several other occasions.

Dirac's view was that one should never state something as true unless one is certain that it is indeed true. He also used this in a humorous form.

Pauli and Dirac were riding in a train through the countryside. After a while, in order to break the silence, Pauli looked for something to say. He saw some sheep and commented, "It looks like the sheep have been freshly shorn." Dirac looked out the window for some time before he replied, "At least on this side."

Dirac's penchant for precision is illustrated even better by the following joke, which was one that he told many times with great joy. A new priest in a small village on his first tour of his parish visits a somewhat modest home overflowing with small children. To his inquiry as to how many children the couple has, the reply is, "Ten, five pairs of twins." The astonished priest asks, "Have you always had twins?" The lady of the house replies, "No Father, sometimes we had nothing."

Here are several more stories that illustrate Dirac's precision.

On a walk with Dirac around a lake, the physicist Behram Kursunoglu saw some birds on the water and counted fourteen of them. When he stated this number to Dirac, the latter replied, "No, there are fifteen. I saw one dive."

At the home of Niels Bohr in the 1950s Dirac was studying an abstract painting that consisted of white dots and little red triangles that were supposed to represent the bodies of chickens and their combs. A lady next to Dirac asked him for his opinion of this painting. His response was, "How many chickens do you think there are there?" Nicholas Kemmer (1911 – 1998), standing next to Dirac, took up this challenge, counted quickly and replied, "Nineteen." Dirac responded, "Eighteen."

Kemmer counted quickly once more and again came up with nineteen. When Dirac still insisted on eighteen they decided to count together. They proceeded to do so and after reaching eighteen, Kemmer pointed triumphantly to a white spot in the corner, "Nineteen." There was, however, no red triangle and Dirac responded quietly, "That one might be a pigeon."

H. R. Hulme, on a walk with Dirac, apologized for the rattle in his pocket by explaining that a bottle of pills he was carrying was no longer full. Dirac responded by noting, "I suppose that it makes the maximum noise when it is half full."

Also at tea time in Cambridge someone commented that recently in Cambridge a disproportionately large number of girls were born. Dirac explained, "It must be something in the air." After a pause he added, "or something in the water."

Dirac was usually difficult to approach while in Cambridge. A colleague is supposed to have asked Dirac, who was sitting beside him during dinner, what he was working on. Dirac asked, "Do you know what adiabatic invariants are?"

"No."

"What then is the use of my talking to you if you don't know the very elements of the subject?"

In 1955, when Jagdish Mehra was still a young man and had just returned to England after spending two years with Heisenberg, a friend of his arranged to take him to dine at the High Table in St. John's College so that he might see Dirac who often ate there. By good fortune Dirac was there and Mehra even got to sit beside him. Somewhat tongue-tied in the presence of this great man, Mehra remembered that it is always good form in England to talk about the weather. Since the weather was very blustery outside, Mehra turned to Dirac with the words, "It's very windy to-day Professor Dirac." Dirac finished chewing and without a word got up and left. Poor Mehra could not guess how he might have offended Dirac. The latter merely went to the door, opened it, looked out, returned to his chair and responded, "Yes."

Fifteen years later Mehra reminded Dirac of this incident. After some thought Dirac said, "I wonder why I did that, because I must have already known that it was windy outside, unless the weather changed since I entered."

Bohr usually needed someone as a sounding board to sharpen his ideas. On one occasion he was working with Dirac and paced up and down while dictating a letter with Dirac writing it down. In the middle of a sentence he stopped and turning to Dirac said, "I don't know how to finish this sentence." Dirac put the pen down and responded, "I was taught at school that you should never start a sentence without knowing its end."

Dirac also always thought before speaking. Sir John D. Cockroft (1897 – 1967) who, together with Ernest T. S. Walton (1903 – 1995), designed one of the earliest particle accelerators was an engineer, as Dirac had also started out. One evening during the soup course at the head dinner table he shouted to Dirac, "Do you consider yourself to be an educated man, Dirac?"

No response. At the end of the third course Dirac finally replied softly, "No. I don't know any Latin or Greek."

Regarding the paranormal, there is a single recorded statement by Dirac. He was at a meeting in a castle, when one of the guests remarked that a certain room was haunted: at midnight a ghost was said to appear. With his usual precision Dirac asked, "Is that midnight Greenwich time, or daylight saving time?"

Dirac's power of analysis extended beyond physics. On a visit to the home of Sir Rudolf Peierls, Dirac watched closely all evening as Mrs. Peierls knitted. At the end of the evening as they were leaving he said to her, "There is a topologically inequivalent way of doing what you were doing." After she asked him what he meant and he showed her she said, "Yes, that's called purling."

As a teacher, Dirac was not one to waste energy. The colleges of Cambridge set fresh exams every year for admission. During the war Dirac was asked to set some questions. The following year he was asked to serve again and to everyone's surprise he set the same questions as in the previous year. When asked about this he replied, "Well, they were good questions last year and they will be good questions this year."

In 1970 Eugene P. Wigner called B.N. Kursunoglu to make arrangements about his visit to Miami. Kursunoglu informed him that his "famous bother-in-law"—referring to Dirac—would be there to meet him at the airport. Wigner replied that this was totally unnecessary since, "I came all the way from Hungary and did not get lost. Why should I get lost in the Miami airport?" When Kursunoglu related this to Dirac, the latter replied, "When Wigner came from Hungary he was young and had much time. But when he comes to the airport next week, he won't have that much time."

This is how Dirac taught Mrs. Kursunoglu about symmetry. "This is easy to remember, just watch the symmetry. When a man says 'yes' he means 'perhaps'; when he says 'perhaps' he means 'no'; when he says 'no' he is no diplomat. When a lady says 'no' she means 'perhaps'; when she says 'perhaps' she means 'yes'; when she says 'yes' she is no lady."

Also at a conference in Miami, a Mr. Gusman, a very wealthy, self-made Cleveland banker met Dirac and told him, "I never sell stocks." To which Dirac replied, "I never buy stocks."

In a reception line, in Miami, a socialite shook Dirac's hand and gushed, "Oh, Professor Dirac, what does it feel like to be the most brilliant man in the world?" Dirac stared at her in silence. After a few minutes she giggled nervously and moved down the reception line. Dirac turned to the person beside him and asked, "What can you possibly say to someone like that?"

With regard to the time shortly after quantum mechanics was completed Dirac commented, "It was a time when second rate physicists could do first rate work." However, he also stated that, "One must not judge a man's worth from his poorer work; one must always judge him by the best he has done."

That Dirac was shy was well known, but some tales are slightly exaggerated. Shortly after Dirac married Wigner's sister a student came one evening to see him. An attractive young woman in a housecoat answered the knock on the door and a seemingly very flustered Dirac rushed over to make the following introduction, "Allow me to present Wigner's sister."

I asked Mrs. Dirac at a conference in Coral Gables in 1969 if the story were true. Her response was, "Partly. What he said was, 'Allow me to present Wigner's sister, who is now my wife'."

Here is a poem written by Dirac while still a student.

> Age is of course a fever chill
> That every physicist must fear
> He's better dead than living still
> When once he's passed his thirtieth year.

Josef Maria Jauch (1914 – 1974) wrote a small popular article with the title *Are Quanta Real?* It dealt with the difficulties of the interpretations of quantum mechanics and Jauch happened to finish it when both he and Dirac were visiting the university in Tallahassee. Anxious to benefit from Dirac's advice he left the article with him. About a week later Dirac returned the article and in his usual taciturn manner simply stated, "Thank you." Anxious to get more than that, Jauch persisted and asked him what he thought of the article. The reply was, "I don't like the title."

"Why not?"

"Well, it's like asking, is God real?"

Jauch was pleased with this response and went on, "That's very interesting, because that is exactly what I wanted to get across."

To this Dirac replied, "Why then did it take you so many pages to say it?"

The following story was related to me by Professor Herbert (Bert) S. Green (1920 – 1929) when he visited Edmonton. Bert Green was Max Born's last student.

On a visit to Adelaide, Australia, Dirac stayed with the governor of South Australia, Marcus L. E. Oliphant (1901 – 2000), at government house. Mark Oliphant had been a student of Rutherford, and built the first high-energy accelerator in England.

When Dirac visited Adelaide, it was winter there. Despite this, Dirac wanted to go swimming since this, besides walking, was one of his great pleasures. This led to the following experimental procedure. One morning Professor Green of the University of Adelaide received a telephone call from Oliphant. "Dirac wants to go swimming and there's no way I'll go. So, you're elected to go with him."

Green, who also enjoyed swimming, but at higher temperatures, had little choice and went to pick up Dirac. Mrs. Dirac, always concerned about her husband's welfare handed Dirac a thermometer with the words, "No swimming if the water temperature is below 15 degrees Celsius."

Dirac, being an attentive husband and a good experimentalist kept the thermometer carefully warm inside his jacket. Once at the beach he removed it quickly, dipped it once into the water and immediately took a reading. To his great delight, the temperature was well above 15°C.

Oliphant, had demonstrated his executive abilities while head of the physics department in Birmingham. In the tea room at Birmingham the faculty were standing around one day and discussing physics. Now it is a fact that physicists, especially theoretical physicists, are incapable of talking about their subject without a blackboard or something else to write on. Oliphant put down his cup and stated, "We need a blackboard in here. You can't properly discuss physics without a blackboard." An assistant was dispatched to find one. Soon he returned with the news that he had indeed found a blackboard, but that it was too big to fit through the door. Unperturbed, Oliphant said that it would fit through the window. A block and tackle were found and twenty minutes later a blackboard was installed in the tea room. Elsewhere bureaucracy would have required numerous forms and several days or weeks of delay.

On his 1955 visit to India to deliver the Rutherford memorial Lecture, Oliphant was first invited to Madras. He had planned to deliver the same lecture in both Madras and at the Raman Insitute in Bangalore. A reporter from the Madras paper *The Hindu*, asked to publish the gist of the lecture and Oliphant gave him a copy of the complete lecture. When he later arrived in Bangalore, Raman greeted him with enthusiasm and told him how much he had enjoyed reading the lecture he presented in Madras. The embarrassed Oliphant had to quickly prepare another version of the same lecture to avoid some biting comments from Raman.

Dirac also visited the Raman Insitute. Now, Raman had a different (incorrect) theory from the Born von Kàrman theory[1] of lattice dynamics and criticized Born and Debye fiercely in public. When Dirac visited Raman's Institute, Raman subjected him to a detailed exposition of his theory and asked his opinion about it all. In his usual careful manner Dirac started

[1] The Born von Kàrman theory is a modification of the Debye theory of specific heat of solids.

hesitantly and said, "What you presented appears reasonable." Before he could continue Raman took his hand, shook it warmly and said, "I know you will see my point. You are one of the greatest physicists for whom I have a great regard."

There is a rather poignant story, that came out many years later at a celebration of Dirac's seventieth birthday. It regards the Schrödinger equation and quantum mechanics. Before Schrödinger had come up with his equation one, Cornelius Lanczos (1893 – 1974), had published a paper in which he also had a wave equation, written in an unusual manner as an integral equation instead of the more familiar form as a differential equation. This equation was equivalent to Schrödinger's equation. His paper also contained some of the material of matrix mechanics. However in integral equations the interesting quantity—in this case the energy—appears as a reciprocal rather than directly. This was cause enough for Pauli to condemn the result publicly as well as in a letter to Pascual Jordan in which he wrote, "On the whole I believe that Lanczos' approach is not very useful." As a consequence Lanczos switched to Relativity and Mathematical Physics, areas in which he became famous and never published another paper on quantum mechanics.

In 1972, at the Trieste Conference to celebrate Dirac's seventieth birthday, Bartel Leendert van der Waerden (1903 – 1996), unaware that Lanczos was in the audience, related this story to an assembled audience of physicists. When Léon Rosenfeld (1904 – 1974), the chairman of the session pointed out that, "We are happy that Professor Cornelius Lanczos is present here and has heard this well deserved vindication of his contribution at that time", a slight, elderly man with long white locks stood up. Rosenfeld continued, "I remember having read your paper, Professor Lanczos, but I cannot claim that I understood it so well as Professor van der Waerden has explained it." At this point Lanczos was invited onto the podium and the following exchange took place.

B. L. van der Waerden, "Oh, You are Lanczos?"

C. Lanczos, "Yes."

B. L. van der Waerden, "Oh, that is marvelous. I didn't know that you were here at this symposium or that you would come to this lecture."

After this there were some questions and van der Waerden handed the microphone to Lanczos. "Now that you have this instrument in your hand, may I ask you a question? Did you know all this to which I have referred in my paper? Were you aware of these connections?"

The answer was, totally out of keeping with this mild-mannered old gentleman, "You are absolutely right. You rehabilitated my work. Pauli was a vicious man, as everyone knows. Anything, which did not agree with his ideas, was wrong, and anything was right only if he made it, if he discovered it, which is alright for such a great man. He could allow himself such viciousness, but I am very grateful to you for pointing out what you have."

Here is a comment by Lanczos. "Most of the arts, as painting, sculpture, and music, have emotional appeal to the general public. This is because these arts can be experienced by some one or more of our senses. Such is not true of the art of mathematics; this art can be appreciated only by mathematicians, and to become a mathematician requires a long period of intensive training. The community of mathematicians is similar to an imaginary community of musical composers whose only satisfaction is obtained by the interchange among themselves of the musical scores they compose."

Although at the 1972 Trieste Conference Lanczos was already quite frail and only had two more years to live, I recall that at the banquet he danced almost every dance with the pretty institute secretaries.

Dirac retired from the Lucasian chair (originally held by Newton) of mathematics at Cambridge in 1969 and moved with his family to Florida where he held visiting appointments at the University of Miami and Florida State University. He continued to do research right up until his death on October 20, 1984 in Tallahassee, Florida. In 1995 he was honored with a plaque, next to the grave of Newton in Westminster Abbey.

As a final comment on Dirac I quote from Niels Bohr. "Of all physicists, Dirac has the purest soul."

The banquet at the Symposium On the Development of THE PHYSICISTS'
CONCEPTION OF NATURE at the International Centre for Theoretical Physics,
Miramare, Trieste, Italy — Dedicated to P.A.M. Dirac on his seventieth birthday.
Right to left: W. Heisenberg, C.P. Snow, P.A.M. Dirac, E. Wigner.

Chapter 14

Quantum Mechanics is Complete

"Nevertheless I repeat once more that I do not mean to bind myself to these; for in them as in other things I am certain of my way but not certain of my position." Sir Francis Bacon

Shortly after the completion of matrix mechanics Schrödinger, in the sequence of events already discussed, produced his wave theory. It was as different as could be from the matrix theory. Matrix mechanics dealt exclusively with discrete, discontinuous quantities whereas wave mechanics dealt with waves, which are intrinsically continuous. For this reason most physicists, steeped in their classical traditions and conservative by nature, found Schrödinger's theory more appealing.

In spite of their totally different appearances, the two theories: matrix mechanics and wave mechanics, were identical. Erwin Schrödinger himself completed the formal mathematical construction of the theory of quantum mechanics in 1926 when he published a paper that showed the connection between his wave mechanics and the matrix mechanics of Born and Heisenberg. Interestingly, Wolfgang Pauli had produced the same proof somewhat earlier, but had not submitted it for publication. The situation for the two types of quantum mechanics was analogous to having the same story written in two different languages such as English and Chinese. The two versions appear to be totally different, but can be translated back and forth and are identical. Each version has advantages and disadvantages. This is the case with Matrix and Wave Mechanics and that is the reason they both are subsumed under the title Quantum Mechanics. Henceforth, there was only quantum mechanics; the previous theories, matrix mechanics and wave mechanics, were but different versions of the same thing. This was even more

175

obvious in the way Dirac had formulated his version of quantum mechanics.

Even so, Schrödinger had no use for the discontinuous quantum mechanics. He wrote, "This difficult method seemed depressing if not revolting ... It was devoid of any clarity." Also, in the paper in which he showed their equivalence, he had a footnote, "My theory was inspired by L. de Broglie ... and by short, but incomplete remarks by A. Einstein ... No generic relationship, whatsoever with Heisenberg, is known to me. I knew of his theory, of course, but felt discouraged, not to say repelled, by the methods of transcendental algebra, which appeared difficult to me, and by the lack of visualizability."

Similarly, Heisenberg wrote to Pauli regarding Schrödinger's theory. "The more I think about the physical portion of the Schrödinger theory, the more repulsive I find it ... What Schrödinger writes about the visualizability of his theory is 'probably not quite right' in other words it's crap."

The problem was that Quantum Mechanics was not yet a complete theory. Two tasks remained:

1. A rigorous mathematical formulation had to be found.

2. After that the more difficult task remained—an elucidation of the physical meaning of the mathematical operations and symbols involved.

In 1926, Volume 23 of the *Handbuch der Physik*, (sometimes referred to as the "Blue Bible") appeared. It contained an article by Pauli on what is now called the Old Quantum Theory. In 1933 in Volume 24, Pauli had an article on the New Quantum Mechanics. Of this article he said, "not quite as good as my first Handbook article[1], but in any case better than any other presentation of quantum mechanics."

The problem of a mathematical formulation was solved by two separate schools. In Göttingen, David Hilbert, the leading mathematician of his time (and most other times) had been consulted all along on the mathematical problems that arose in the formulation of quantum mechanics. Thus, he was well-versed in the current status of quantum theory and picked up Heisenberg's challenge to pure mathematicians by giving a series of lectures on the subject. One of his most brilliant disciples was Johann (Johnny) von Neumann who in 1932 published *Mathematical Foundations of Quantum Mechanics* that laid down a completely rigorous mathematical framework for quantum theory.

In the meantime, unimpeded by conventional details of mathematical

---

[1] Pauli's first *Handbuch* article was the one on General Relativity that he wrote while still a student with Sommerfeld.

rigor and guided by a sure physical intuition Dirac had built, in a series of brilliant papers, an alternate mathematical framework that was far more appealing to physicists. The famous mathematician, Harish-Chandra (1923 – 1983), was Dirac's graduate student. The following exchange between him and Dirac, while working on the representations of the Lorentz group, may have contributed to his becoming a mathematician and illustrates Dirac's approach. When Harish-Chandra complained, "My proofs are not rigorous", Dirac replied, "I am not interested in proofs, but only in what nature does."

The second, more difficult task, was the physical interpretation of the symbols involved. From the start, Schrödinger tried very hard to give his wave theory a nice physical interpretation in keeping with classical ideas. He believed that the intensity of his psi waves gave the charge density of the particles being described. His conjecture was supported by the fact that he was able, in one instance, to use his wave theory to construct wave packets or "lumps" that were concentrated in small regions of space and thus approximated small particles quite well. However, as Heisenberg soon showed this was a fortuitous coincidence since Schrödinger happened to choose very particular wave packets for which this happened to be true. In general, wave packets spread out very quickly to macroscopic dimensions and could not possibly correspond to small particles.

For a while Schrödinger seemed to have become less attached to his models. In 1928 he stated in a lecture that, "We must not forget that pictures and models finally have no other purpose than to serve as a framework for all the observations that are in principle possible." Even later in life he admitted, "I would always have the electron in mind as a small ball, as a sphere. I would say that it was also useful sometimes to call it a wave, but only as a way of talking, not as reality."

De Broglie also did not think of Schrödinger's psi waves as being real. He named them "pilot waves". By this he meant that in some sense, the psi waves determined the classical path of the particle. In his own words, "I especially placed the particle into the continuous wave and assumed that the propagation of the wave carried the particle with it. The psi wave in a sense 'shows the way' to the traveling particle."

Now came the final phase, the interpretation of this quantum mechanical formalism. Max Born made the initial break-through, but the major interpreters of the new gospel were Niels Bohr and Werner Heisenberg challenged and goaded by Albert Einstein. The first two played the role of the angels whereas Einstein played the unwilling role of devil's advocate.

Max Born, who disagreed strongly with Schrödinger's attempts at a continuum interpretation without quantum jumps, found the interpretation of the Schrödinger waves, now most commonly accepted. In Born's own words, "On this point I could not follow him. This was connected with the fact that my institute and that of James Franck were housed in the same building of the Göttingen University. Every experiment by Franck and his assistants on electron collisions (of the first and second kind) appeared to me as new proof of the corpuscular nature of the electron."

In 1926 Born repeated, "It is necessary to drop completely the physical pictures of Schrödinger which aim at a revitalization of the classical continuum theory, to retain only the formalism and to fill that with new physical content." And again, "The classical theory introduces the microscopic coordinates which determine the individual processes only to eliminate them because of ignorance by averaging over their values; whereas the new theory gets the same results without introducing them at all."

The experiment by James Franck that Born refers to is the famous Franck-Hertz experiment. The Gustav Hertz (1887 – 1975) in this experiment is the grand-nephew of the Heinrich Hertz who had discovered the photoelectric effect. Besides being a good physicist, Gustav Hertz had a strange sense of humor. He occasionally came to drink tea with the chemists in the laboratory where Otto Hahn (1879 – 1968) and Lise Meitner (1878 – 1968) worked. One time he refused the tea stating, "This is too bland, hand me the alcohol," and he had one of the students reach for a bottle of pure alcohol. Lise Meitner was horrified, "But Hertz, you can't drink that, it's poison." Ignoring her, Franck poured himself a large tumbler full and drained it with no effect. He had earlier got the student to fill the bottle with pure water.

Born proceeded to try and explain the above-mentioned collision processes by using Schrödinger's wave theory. After he had finished this paper on the collision of electrons and hydrogen atoms he gave it to Robert Oppenheimer, who was then studying with him at Göttingen. Born wanted him to check the involved calculations. When Oppenheimer returned the manuscript he stated, "I couldn't find any mistake. Did you really do this alone?" The astonishment expressed by these words and clearly visible on his face was excusable since Born was notorious among his students for not being able to carry out long computations without making silly mistakes. Oppenheimer, however, was the only one frank and rude enough to say so without joking.

The following story told me by Avadh Bhatia (1921 – 1984) is another example of Born's tendency to make computational errors. In the 1940's when Bhatia and R. B. Dingle were postdoctoral fellows with Born, he gave them one of his manuscripts to review. These two young physicists checked his work and found several silly mistakes. The next day when Born asked them about the work they pointed out that they had found some mistakes. Much to their surprise, Born became quite angry and even threw chalk at them while declaring that they were not competent to understand his work. Naturally, both Bhatia and Dingle were visibly distraught and doubted whether they had the necessary ability to do physics. The next day Born returned, apologized, and told them that they were right. Both young physicists went on to have brilliant careers: Bhatia as a condensed matter physicist and Dingle as an applied mathematician specializing in asymptotic expansions.

For Oppenheimer's final oral examination for the Ph.D. degree, James Franck was one of the people that attended. After the examination someone asked him how Oppenheimer had done. He answered, "I don't know. I managed to sneak out when he started asking us questions."

Born's collision calculation, that Oppenheimer had checked, was highly successful, but Born had been forced to make a radical interpretation. "The intensity of the wave at any point gives the probability for finding the particle at that point." This was, in effect, an actualization of Einstein's old concept that somehow light quanta or photons were guided in their motion by the electromagnetic waves of Maxwell's equations. Born soon realized that the probability concept he had introduced differed radically from the probabilities to which physicists were used.

Normally, if one has two events with separate probabilities, then the probability of observing either one or the other or both is just the sum of their separate probabilities. Thus, if you flip two coins the probability of observing two heads is one quarter; the probability of observing two tails is also one quarter and the probability of observing either two heads or two tails is one quarter plus one quarter or one half. Not so for quantum mechanical probabilities! The result may be bigger or smaller than the sum because here one does not add probabilities (the intensities of the Schrödinger waves) instead one adds the two waves directly and computes the probability (intensity) for this compound wave to get the probability. This result differs dramatically from the classical probability theory. As is well known from sound waves and all other waves, the two waves can reinforce each other to produce a wave with four times rather than just twice the intensity or else the waves can interfere with each other to produce zero intensity.

This interpretation by Born seemed to make any reconciliation of quantum theory with classical theory impossible. Nevertheless both Einstein and Schrödinger continued for the rest of their lives to try to remove this "unattractive" feature. In explaining his interpretation Born said, "We free forces of their classical duty of determining directly the motion of particles and allow them instead to determine the probability of states. Whereas before it was our purpose to make these two definitions of force equivalent, this problem has no longer, strictly speaking any sense. The only question is why the classical definition is so successful for such a large class of phenomena. As always in such cases, the answer is: Because the classical theory is a limiting case of the new one."

Born emphasized how classical mechanics was not invalidated by either relativity or quantum mechanics, but was simply an approximation to these newer theories. "And the continuity of our science has not been affected by all these turbulent happenings, as the older theories have always been included as limiting cases in the new ones."

Furthermore, Born emphasized the quantum nature of phenomena and the impossibility of following in detail a transition, as for instance between two Bohr orbits in the Old Quantum Theory. Again in his words, "Whatever occurs during the transition can hardly be described within the conceptual framework of Bohr's theory, nay, probably in no language, which lends itself to visualizability."

Most physicists today would still agree with Born. Of course, Born had not solved the problem of interpretation completely; he had, however, shown the way. Certainly it was important to know that Schrödinger's waves gave probability amplitudes, but what about dynamical variables like energy, momentum, position, etc.? Also Schrödinger's interpretations, as well as at least two other ones, one by Count Louis-Victor de Broglie, were available. How to choose between them? The first major skirmish between the Schrödinger and Born interpretation was fought at Niels Bohr's institute in Copenhagen in September 1926 when Schrödinger visited there to lecture on wave mechanics. Heisenberg was also present as a consequence of the following circumstances.

As a student, Heisenberg had almost failed his final oral examination because at that time he was not very interested in experimental work and had prepared in it so badly that Wilhelm Wien noticed this. Consequently Wien asked him several detailed questions about experimental technique. When Heisenberg failed to answer these satisfactorily, Wien became angry and declared that Heisenberg had failed. Sommerfeld knew that Heisenberg was his best student and his thesis a brilliant piece of work. After a long and heated discussion, Sommerfeld succeeded in procuring a pass for Heisenberg.

The irony is that one of Wien's questions concerned the resolving power of a microscope and Heisenberg was so conscientious that after the exam he researched all the questions asked, including this one. This is why, later in his interpretation of quantum mechanics, he was able to give a physical argument, not only a mathematical one for his uncertainty relation

$$\Delta x \Delta p \geq h/4\pi$$

in what is now often referred to as the Heisenberg microscope. This may also be why Heisenberg later defined an expert as follows. "An expert is someone who knows some of the worst mistakes that can be made in a subject and how to avoid them." There is a similar statement by Bohr, "An expert is a man who has made all the mistakes, which can be made, in a very narrow field."

Schrödinger and Heisenberg met for the first time, by coincidence, late in the summer of 1926 at a theoretical seminar in Munich organized by Sommerfeld. Heisenberg happened to be in the audience since he was spending part of his vacation at his parents' house. Wilhelm Wien, Director of the Experimentalinstitut, was also in the audience. He had a strong dislike for matrix mechanics, but was not unhappy with the wave mechanics. Heisenberg did not hesitate to criticize Schrödinger's overly optimistic view of wave packets. This was too much for Wien, who remembered that only three years earlier he had almost failed this student for his inability to answer his question regarding the resolving power of the microscope. Now this young upstart was criticizing a professor as if he were his equal. Forgetting his dignity, Wien jumped from his seat and addressed the former student, "Young man, you have yet to learn physics and it would be better if you resumed your seat." In a milder tone he stated that the feelings of the poorly educated young man were understandable since wave mechanics put an end to all this nonsense of quantum jumps. Regarding the issues he was raising he felt that, "We do not doubt that Herr Schrödinger will overcome them in the nearest future."

Actually, H. A. Lorentz had already presented these same objections to Schrödinger in a letter of May 1926. "But a wave packet can never stay together and remain confined to a small volume in the long run . . . Because of this unavoidable blurring, a wave packet does not seem to me to be very suitable for representing things to which we want to ascribe a rather permanent individual existence."

Heisenberg was depressed after this meeting with Schrödinger and wrote to Bohr, who as a consequence invited both Schrödinger and Heisenberg to Copenhagen.

The three main creators of quantum mechanics in the Stockholm train station. From left to right: Heisenberg's mother, Schrödinger's wife, Dirac's mother, Paul Dirac, Werner Heisenberg, Erwin Schrödinger.

As has been repeatedly emphasized, Schrödinger had no love for these quantum jumps and the accompanying statistical interpretation. He wanted a theory along the lines of classical physics. Thus, he launched a fierce attack on Bohr's concepts of quantum jumps and discontinuity. Bohr defended himself so ably that finally Schrödinger had to retreat and exclaim in exasperation, "If one has to stick to this damned quantum jumping then I regret ever having been involved in this thing." To this Bohr replied, "But we others are very grateful to you that you were, since your work did so much to promote this theory."

The conflict continued from morning till night. Finally, Schrödinger, always of a frail constitution, fell ill and had to be confined to bed in one of the upstairs bedrooms of the Carlsberg mansion. Even so, Bohr could not desist and visited him up there regularly. Other visitors to the mansion often heard Bohr's exclamation float down from upstairs, "But Schrödinger!" and so the debate continued.

Schrödinger wrote in response to the discontinuous quantum jumps and Bohr's idea of jumping between stationary states. "A theoretical science, unaware that those of its constructs considered relevant and momentous are destined eventually to be framed in concepts and words that have a grip on the educated community and become part and parcel of the general world picture—a theoretical science, I say, where this is forgotten, and where the initiated continue musing to each other in terms that are, at best, understood by a close group of fellow travellers, will necessarily be cut off from the rest of cultural mankind; in the long run it is bound to atrophy and ossify."

In July 1927 Schrödinger, in a letter to Planck regarding the statistical interpretation, confessed, "I would be willing to believe it since the interpretation is really much more convenient, if I could only pacify my conscience and convince it that it is not frivolous to get off so easily in overcoming the difficulties." In 1945 he told his student, H. W. Peng, "Quantum mechanics was born in statistics and it will end in statistics." Later, he again wrote, "If you cannot, in the long run, tell everyone what you have been doing, your doing was worthless."

Heisenberg would have agreed with this last statement since he himself said, "Even for the physicist the description in plain language will be a criterion of the degree of understanding that has been reached." On the other hand he also stated something with which most physicists would now agree, "If we want to describe what happens in an atomic event, we have to realize that the word 'happens' can apply only to the observation, not to the state of affairs between two observations." He also wrote, "Natural science does not simply describe and explain nature; it is part of the interplay between nature and ourselves; it describes nature as exposed to our method of questioning." This is very different from the classical viewpoint which assumes that there exists an absolute reality out there and that science attempts to uncover the details of this reality.

Schrödinger's visit to Copenhagen provided the impetus for vigorous discussions for many weeks. The biggest impact, however, was on Bohr and Heisenberg who recognized the difficulties for what they were: a lack of interpretation of the theory. Day and night these two tossed ideas back and forth until, after several months, they were both exhausted. As Heisenberg recalled, "Neither of us could understand how to reconcile such elementary phenomena as the electron path in a cloud chamber. Our discussions frequently continued long past midnight. Our strenuous efforts did not yield any results and after a few months we were totally exhausted and our nerves were stretched to the limit." At this stage they decided to go for a ski vacation to Norway. However, the next morning Bohr had left by himself. Later Heisenberg explained, "He wanted to be by himself, and I think he was quite right. On the whole, I was glad that he left me alone in Copenhagen where now I had a chance to think quietly about these hopelessly complicated problems. I focussed my efforts on making a mathematical description of the electron path in the cloud chamber. Quite soon I satisfied myself that the accompanying difficulties were quite insurmountable and I started thinking whether all questions had been wrong. But, where had we gone astray?"

During Bohr's absence, Heisenberg continued to try to reconcile the fact that his darling, matrix mechanics, gave only discontinuous values and yet in cloud chambers one could clearly see the tracks of electrons. How was this possible? He was sure that the mathematical formalism was correct. He even stated, "Mathematics is clever enough to do everything by itself without physicists' speculation."

He recalled that when he had told Einstein that in developing matrix mechanics he had tried to follow Einstein and construct a theory that only included observable quantities just as Einstein had in developing relativity. "But, I only used the same principle as you." To this Einstein had replied, "I may have used such a principle once upon a time, but it is wrong. It is like a joke which may be fine once but is usually not good the second time." Einstein had then elaborated and told him that a theory need not only contain observable quantities, but that the theory must decide what is observable. This put Heisenberg onto a new line of thought.

He now asked himself a series of questions. What kind of experiments are possible in principle? How is it that cloud chamber tracks are possible? He finally realized that a cloud chamber track is not an exact picture of the trajectory of a particle. The individual droplets that make the track visible are of a certain size and that limits the accuracy with which the trajectory can be measured. So, he again asked himself something like, if I measure the exact position of a particle with perfect equipment, can I simultaneously measure its momentum with arbitrary accuracy? Consideration of these questions led Heisenberg to formulate what is now called the Indeterminacy or *Uncertainty Principle*. Simply stated it says that the answer to the second question above is, "No!"

More generally, the Uncertainty Principle states that it is impossible, even in totally idealized situations, to devise experiments that allow one to measure both of certain pairs of dynamical variables (called conjugate variables) with arbitrary accuracy. The most well-known pair of such conjugate variables is position and momentum. Dirac and Jordan, on purely mathematical grounds, had already anticipated some of these results, but Heisenberg investigated and elucidated much more fully their physical implications. When Bohr returned from his ski trip, Heisenberg showed him his results. Bohr was delighted.

Heisenberg's investigation culminated in a series of, "thought-experiments". These are experiments not actually performed, but simply carried out in the physicist's imagination with perfect equipment. One of the best known of these is the Heisenberg Microscope used to "measure" both the position and momentum of a particle to the maximum accuracy possible. To measure the position of a particle as accurately as possible requires using light of the shortest possible wavelength. This limits the accuracy of such an idealized measurement to roughly the wavelength of the light used. So, the shorter the wavelength of the light, the more accurate the measurement of the position.

On the other hand, the de Broglie relation shows that the shorter the wavelength of the light, the greater the momentum it carries and hence the greater the recoil of the particle illuminated by this photon of light. This "kick" or recoil is what Compton measured and is known as the Compton effect. The upshot of this analysis is that the more accurately we try to determine the location of the particle, the bigger a kick we have to impart to it so that the less accurate it becomes to know its momentum. The result again involves Planck's constant. The precise statement of this result is that the inaccuracy in position multiplied by the inaccuracy in momentum is greater than or equal to Planck's constant divided by 4π.

Finally Heisenberg was able to state, "Quantum mechanics is a complete theory in the sense that information more detailed than that permitted by the Uncertainty Principle can not be obtained under any circumstances."

By the way, "Uncertainty Principle" is a not too accurate translation of Heisenberg's original word *Unbestimmtheitsprinzip*. A more accurate translation would be "Indeterminacy Principle". But could we then make such jokes as, "Heisenberg might have been here, but we're not certain"?

The important point that followed from the Uncertainty Principle is that the mechanistic all-predictive world of the nineteenth century physicist was destroyed. Ever since Pierre Simon de Laplace (1749 – 1827) had hypothesized, in his *Mécanique Céleste*, that it was possible, in principle, to predict completely the precise evolution of the universe if one knew the exact position and momentum of every particle in the universe at one instant of time, most classical physicists had believed in such a mechanistic view.

Napoleon is said to have also read *Mécanique Céleste* and after doing so, reputedly questioned Laplace on his neglect to mention God. Laplace replied that he had no need for that hypothesis.

Heisenberg, however, showed that such knowledge of the exact position and momentum of every particle in the universe at one instant of time was impossible even in principle. Philosophers, who had worried how free will was possible in Laplace's mechanistic universe, breathed a collective sigh of relief. Physicists, being more pragmatic, knew that this did not solve the problem of free will, since no connection between mind and matter had been established by these cogitations.

Actually, Max Born went even further. In *My Life and Views* he wrote, "The deterministic interpretation of Newtonian mechanics is actually an unjustified idealization, as Brillouin[2] and I have independently shown. It is based on the idea of absolutely precise measurements, an assumption which has no physical meaning. It is not difficult to write classical mechanics in a statistical form." Born was aware that very high accuracy would not suffice since there are many systems that show dramatically different behavior for extremely tiny changes in initial conditions. This is the basis of the modern Theory of Chaos.

During his Norwegian ski trip, Bohr's powerful mind had also been struggling with the attempt to reconcile the various conflicting aspects of quantum mechanics and to mould them into a logically consistent whole. He returned with a solution. Thus, on the one hand, particles could be observed. Yet, these very same particles could be made to display all the characteristics of waves. To resolve this dichotomy Bohr developed the concept of complementarity. The idea is that objects in nature possess conflicting or complementary properties. The accurate observation of one of these properties precludes the possibility of observing the other. Yet both properties are necessary for a complete understanding of the object. Heisenberg's Uncertainty Principle provided, in fact, a concrete example

---

[2]Marcel-Louis Brillouin (1854 – 1948)

of Bohr's ideas. Later Bohr extended these ideas to concepts other than physics. In fact he claimed that precision of a statement and clarity of the statement were complementary since the more one tried to be precise the more one was bound to be obscure. He managed to demonstrate this over and over again in his talks because he tried very hard to be precise and thus achieved the complementary result. An example is given below by Bohr's statement on understanding.

While working with Bohr on quantum mechanics Heisenberg asked Bohr, "If the inner structure of the atom is as closed to descriptive accounts as you say, if we really lack a language for dealing with it, how can we ever hope to understand atoms?" After a hesitation Bohr replied, "I think we may yet be able to do so. But in the process we may have to learn what the word 'understanding' really means."

Heisenberg must have thought very much along the same lines because in 1958 he said, "If we want to describe what happens in an atomic event we have to realize that the word 'happens' can apply only to observation, not to the state of affairs between two observations."

On other occasions Bohr made the following statements, "When it comes to atoms, language can be used only as in poetry. The poet too, is not nearly so concerned with describing the facts as with describing images."

At the Carlsberg mansion, one evening, Bohr attempted to explain to the philosopher Harald Høffding the double slit experiment. This experiment illustrates the complementarity principle since it can be used to demonstrate both the wave nature as well as the particle nature of an electron. It is a wave because it produces an interference pattern and it is a particle because it hits the observation screen at a definite point. Someone listening made the remark, "But the electron must be somewhere on its way from the source to the observation screen." To this Bohr replied, "What is in this case the meaning of the word 'to be'?" The philosopher Jørgen Jørgensen, who was also present, protested, "One cannot, damn it, reduce the whole of philosophy to a screen with two holes."

Similarly, after a discussion, someone at Bohr's institute said that it made him quite giddy to think about these questions. Bohr's response was, "But if anybody says he can think about quantum problems without getting giddy, that only shows that he has not understood the first thing about them."

Many years later, Richard Feynman (1918 – 1988) echoed these same sentiments. "A philosopher once said, 'It is necessary for the very existence of science that the same conditions always produce the same results.' Well, they don't!"

Also at Cornell, Richard Feynman, in a lecture The Character of Physical Law, repeated in a similar vein. "I think I can safely say that nobody understands quantum mechanics. So do not take the lecture too seriously, feeling that you really have to understand it in terms of some model that I am going to describe, but just relax and enjoy it. I am going to tell you what nature behaves like. If you will simply admit that maybe she does behave like this, you will find her a delightful, entrancing thing. Do not keep saying to yourself, if you can possibly avoid it, 'But how can it be like that?' because you will go 'down the drain' into a blind alley from which nobody has yet escaped. Nobody knows how it can be like that."

Here are two more Bohr quotes. "It is wrong to think that the task of physics is to find out how nature is. Physics concerns what we can say about nature." Also, "Oh no, I can't believe that. That is much too concrete for it to be real; that is only formal."

Another example of Bohr's inability to be both clear and precise is demonstrated by what happened at the yearly conference in 1932, in Copenhagen. Bohr lectured in his usual unintelligible way. He had a piece of chalk in his right hand and a sponge in his left and he would erase an equation with his left hand almost as quickly as he wrote it with his right hand until his friend Ehrenfest stood up and shouted, "Bohr, give me that sponge!" Bohr handed the sponge over and Ehrenfest kept it for the rest of the lecture.

Even de Broglie referred to Bohr "with his predilection for obscure clarity" as the "Rembrandt of contemporary physics".

These difficulties with Bohr's lectures were due to his attempts to continually "improve" what he was saying. This effort extended to some of his other activities. Thus, it was impossible to play chess in Bohr's presence. Every time a player made a "poor" move Bohr would return the pieces to their original position to "improve" on it.

Knowing about Bohr's need to continue to "improve", it was natural for Pauli to respond, in the following fashion, to an invitation by Bohr to help with criticism of a manuscript, "If the last proof is sent away, then I shall come."

When someone asked Harald Bohr how come he was one of the best lecturers while his brother Niels was so bad he answered, "Simply, because at each place in my lecture I speak only about those things I have explained before, but Niels usually talks about those things which he means to explain later." This is verified by a statement by Niels himself. When Harald and

Niels Bohr went skiing with Richard Courant, Niels injured his knee and was laid up for a few days. During this period he lectured to his brother and friend about quantum mechanics and complementarity. When one of them interrupted him to clarify a question, Niels responded impatiently, "Of course you cannot understand what I try to say now; this may perhaps become understandable, but only after you have heard the story as a whole and have understood the end."

Bohr's insistence on the importance of language for unambiguous communication and that we can really only talk about how things behave is illustrated by his response to Jørgen Kalckar when the latter jokingly objected to Bohr that one might then ascribe consciousness to a computer.

"It is not a question of what one could do, but what is necessary. I am quite prepared to talk of the spiritual life of an electronic computer; to say that it is considering or that it is in a bad mood. What really matters is the unambiguous description of its behavior, which is what we observe. The question of whether the machine really feels, or whether it merely looks as though it did, is absolutely as meaningless as to ask whether light is 'in reality' waves or particles. We must never forget that 'reality' too is a human word just like 'wave' or 'consciousness'. Our task is to learn to use these words correctly—that is unambiguously and consistently."

In the same vein, with regard to the ability of Indian sages to comfort people, he had this to say. "They understood that any use of the word 'meaning' implies a comparison and with what can we compare the whole of existence? The one thing is certain that a statement like 'existence is meaningless' is itself devoid of any meaning."

The opening of the Nuclear Science Institute of the Weizmann Institute in 1958 brought many of the world's greatest physicists together. At the symposium, Niels Bohr gave a long rambling talk that was so difficult to understand that some people even wondered what language he was speaking. At the end of the talk, the chair of the session, S. Chandrasekhar asked if there were any question. Someone asked a question and Bohr rambled on for another half an hour while everyone remained seated. After that no one dared ask another question. Chandrasekhar later commented, "That's the sign of a really great man—people stay for two hours to listen to something they don't understand."

Some further examples of complementarity are:

Cleverness and Wisdom,

Necessity and Freedom (i.e. free will),

Clarity and Truth.

A further illustration of Bohr's approach can be seen from the following quote. "Truth is a statement whose opposite is false. Deep truth is a statement whose opposite also contains a deep truth." As an example of this he noted that both of the following statements contain a deep truth. "There is a God. There is no God."

Even more than thirty years later, Schrödinger refused to accept complementarity, calling it a "thoughtless slogan". He went on to say, "If I were not thoroughly convinced that the man (Bohr) is honest and really believes in the relevance of his—I do not say theory but—sounding word, I should call it wicked." He then quoted two lines from Göthe's Faust of which I give a rather free translation

> For just where concepts you do fail
> Of the proper word you may yourself avail.

Bohr was very much aware of his tendency to illustrate the complementarity of truth and clarity. In this regard, he loved to tell the following story. A young man was sent from his village to another to hear a great rabbi. When he returned he gave the following report. "The rabbi spoke three times. The first talk was brilliant, clear and simple. I understood every word. The second talk was even better, deep and subtle. I didn't understand much, but the rabbi understood all of it. The third talk was a great and unforgettable experience. I understood nothing and the rabbi himself didn't understand much either."

This should be compared with the classification of talks at the Institute for Advanced Studies at the time that von Neumann was there. A talk was simple if half of the audience understood it. A talk was complicated if only 10% of the audience understood it. A talk was truly deep if only the speaker and von Neumann understood it and a talk was earthshaking in its originality and profundity if only von Neumann understood it.

In September 1927, an international congress of physicists convened at Como to commemorate the 100th anniversary of the death of Count Alessandro Volta (1745 – 1827), who had made many fundamental contributions to the study of electricity. The most notable physicist absent was Einstein and so he did not hear Bohr's first presentation of his concept of complementarity. As already stated, this talk was itself an example of Bohr's later extended version of complementarity, namely Clarity and Truth. This presentation which laid the foundations for the later Copenhagen interpretation of quantum mechanics, the interpretation with the largest number of adherents, seems to have made little impact on the physicists present. After the conference, Eugene Wigner remarked that Bohr's talk was certainly not something that would cause any physicist to change his way of doing quantum mechanics. Nevertheless this was the beginning of a battle to establish a new dogma, an interpretation of quantum mechanics. The Como conference may thus be viewed as a preliminary bout, a sort of warm-up for the main event.

The main event took place in October 1929 at the *Fifth Physical Conference of the Solvay Institute in Brussels.* The list of participants at this congress reads like a who's who of physics. Bohr again presented his talk, but this time Einstein was present. Hendrik Antoon Lorentz, the grand old man of Dutch physics, who by now had been passed by the new physics, opened the conference. He stated quite unequivocally, "I wish to have a quite definite picture of any phenomenon. An electron, for me, is a particle which, is at a given point in space at a given time. If the electron collides with an atom, penetrates it and after numerous adventures leaves it, then I visualize a certain line along which the electron travelled in the atom."

This picture, as Lorentz probably realized, is not compatible with quantum mechanics. Later he confessed to Abram F. Ioffe (1880 – 1960), "I have lost the conviction that my research was leading to the objective truth and I do not know what I have lived for. My only regret is that I did not die five years back when everything still seemed clear to me."

Einstein saw this congress as an opportunity to clarify some of his misgivings about quantum mechanics. He had seen Heisenberg's paper on the Uncertainty Principle and was disturbed by the claim that, "Quantum mechanics is a complete theory in the sense that information more detailed than that permitted by the Uncertainty Principle cannot be obtained under any circumstances." In fact, in 1926 he had already stated to Ehrenfest that, "I look upon quantum mechanics with a suspicious admiration." In the same year he repeated in a letter to Born, "Quantum Mechanics is very impressive. But an inner voice tells me that it is not yet the real thing. The theory produces a good deal, but hardly brings us closer to the secret of the Old One. I am, at all events, convinced that He does not play dice." The claim, now repeated by Bohr, that a statistical theory like quantum mechanics gave a description of physical reality, as complete as possible, was contrary to Einstein's whole outlook on science. Such heresy had to be eradicated before it spread. He stated clearly his principle that, *"Der Alte würfelt nicht."* (The Old One does not play dice.)

To this Bohr replied, "Indeed it is no business of ours to tell God how he must govern this world." Twenty years later, in a paper to celebrate

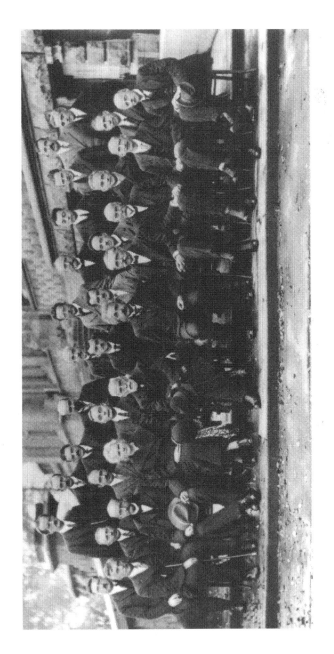

INSTITUT INTERNATIONAL DE PHYSIQUE SOLVAY

CINQUIÈME CONSEIL DE PHYSIQUE — BRUXELLES, 1927

A. PICCARD   E. HENRIOT   P. EHRENFEST   Ed. HERZEN   Th. DE DONDER   E. SCHRÖDINGER   E. VERSCHAFFELT   W. PAULI   W. HEISENBERG   R.H. FOWLER   L. BRILLOUIN

F. DEBYE   M. KNUDSEN   W.L. BRAGG   H.A. KRAMERS   P.A.M. DIRAC   A.H. COMPTON   L. de BROGLIE   M. BORN   N. BOHR

I. LANGMUIR   M. PLANCK   Mme CURIE   H.A. LORENTZ   A. EINSTEIN   P. LANGEVIN   Ch.E. GUYE   C.T.R. WILSON   O.W. RICHARDSON

Absents : Sir W.H. BRAGG, H. DESLANDRES et E. VAN AUBEL

Einstein's seventieth birthday, he recalled his words somewhat differently. "I answered that even the thinkers of antiquity had advocated an extreme caution in attributing to Providence the properties expressed in terms of everyday experience."

Many years later, Einstein's quote inspired the following two quotes.

Jean Untermeyer: "God casts the die, not the dice."

Stephen W. Hawking: "God not only plays dice. He also sometimes throws the dice where they cannot be seen."

During these Bohr-Einstein debates at the Solvay congress Paul Ehrenfest, exasperated with Einstein's stubbornness finally said to his friend, "I am ashamed for you, Einstein, you attack the new quantum theory in just the same way as your enemies attacked the relativity theory." They remained friends in spite of this. Years later when Einstein returned from a trip to Spain, Ehrenfest asked him why he had gone to Spain when there was no physics of interest to him there. The latter replied, "True, but the king gives such nice dinner parties."

The confrontation between these two beliefs occurred throughout the conference: in the corridors, during meals, on the veranda, everywhere. Einstein used his tremendous intellect to devise thought experiments that would allow information more detailed than that permitted by the Uncertainty Principle and thus violate that principle. Bohr fought back. No matter how ingenious Einstein's imaginary experiments, Bohr was always able to detect an essential flaw, namely that at some crucial intermediate stage the Uncertainty Principle was ignored. Only thus were Einstein's thought experiments able to violate this principle. The engagement of the two protagonists was so great that, a year later at the Sixth Solvay Congress, the struggle continued.

Einstein, who had thought about this problem for a year, presented Bohr with the following "experiment" to disprove the time-energy uncertainty relation. He considered a box filled with radiation and with a shutter in one wall. This shutter is operated by a clock, so that it can be opened and shut to release a photon during an accurately determined time interval. Einstein then proposed to weigh the box (with arbitrary accuracy) before and after the photon was emitted. Using his famous formula $E = mc^2$ one would then have the energy and time of the process to arbitrary accuracy. Thus, he claimed, Heisenberg's Uncertainty Principle would be violated.

Bohr spent a sleepless night puzzling over this argument, but by morning he had found the flaw. Einstein had failed to take his own theory of General Relativity into account. The gravitational field clearly affected the clock and when this was computed, the error in the clock rate (determined by the mass of the box) was just such as to again yield the Heisenberg Uncertainty Relation. Einstein's classical deterministic physics was defeated. The Copenhagen Interpretation, as it came to be known, triumphed and henceforth became the dominant interpretation of quantum mechanics. It has remained so to this day, in spite of numerous alternate interpretations, all of which incidentally, also accept the Uncertainty Principle as correct.

That there is still some debate over just what the Copenhagen Interpretation is can be seen from a recent statement by N. David Mermin, a professor of theoretical condensed matter physics at Cornell University. "The Copenhagen interpretation means, 'Shut up and calculate.' "

So in spite of himself, Einstein by forcing Bohr, Heisenberg, and Born to examine the concepts of quantum mechanics in detail, hastened the clarification of this theory. It is nevertheless ironic that Einstein who, back in 1905 with his paper on the photo-electric effect had given such an impetus to the fledgling quantum concept, refused for the rest of his life to accept the theory that finally incorporated so many of his early ideas. As already mentioned earlier, when Einstein listed his major accomplishments he omitted his explanation of the photoelectric effect, for which he had received the Nobel Prize.

The uncertainty principle caused a lot of discussion and continues to do so, both among philosophers and physicists. Thus, Bertrand Russell wrote to Bohr, "I am looking forward to seeing you for various reasons, but among others because I hope to get to understand why Heisenberg's principle is so incompatible with determinism. Left to myself I should only have thought that the things to be determined are not what used to be supposed."

Rutherford also commented to Bohr regarding the uncertainty principle, "Bohr, my boy, you are too complacent about ignorance."

Similarly, Sir James Jeans published in *Nature* in 1934, "After studying Bohr and Heisenberg, nature consists of waves of knowledge or absence of knowledge in our own minds." He also observed that, "The essential fact is that all the pictures which science now draws of nature, and which alone seem capable of according with observational facts, are mathematical pictures."

Born's comment was, "If God has made the world a perfect mechanism, He has at least conceded so much to our imperfect intellect that in order to predict little parts of it, we need not solve innumerable differential equations, but can use dice with fair success."

Thirty-six years later, Werner Heisenberg recalled the electrifying effect of the Fifth Solvay Conference. "I would say that a change had taken place which now I could explain only in terms of lawsuits. That is, 'The burden of proof was reversed'. The burden of proof suddenly went to the Wien people and so on because the rumor spread that there was a group in Copenhagen which can answer every question about experiments. ... So, if you want to do anything against this view you have to disprove them. And the rumor also spread that nobody so far has been able to disprove them, not even Einstein, who does not believe it. Einstein has not been able to disprove these people at the long conference in Brussels. ... Now there were a few people from Copenhagen who told the younger generation, 'You do this and you do that ... now that this is correct and you can go ahead'."

As a young man, Heisenberg had admired Einstein and tried to model himself after him. However, after quantum mechanics became more firmly established he wrote to Einstein in the less respectful tone, "We can console ourselves that the dear Lord God would know the positions of the particles and thus He could let the causality principle continue to have validity."

Heisenberg's whole approach to the meaning of quantum mechanical measurements was summarized by him in the following statement. "What we observe is not nature itself, but nature exposed to our method of questioning."

Regarding new ideas he had this to say. "A really new field of experience will always lead to crystallization of a new system of scientific concepts and laws. When faced with essentially new intellectual challenges, we continually follow the example of Columbus who possessed the courage to leave the known world in an almost insane hope of finding land again beyond the sea."

Einstein still had one last salvo to fire at the quantum theory. In 1935, Albert Einstein, Boris Podolsky (1896 – 1966), and Nathan Rosen (1909 – 1995) published a paper in the *Physical Review* with the title, "Can quantum mechanical description of physical reality be considered complete?" The results of this paper are referred to as the EPR (Einstein, Podolsky, Rosen) paradox.

The paper starts off right away by stating the authors' position. "Any serious consideration of a physical theory must take into account the distinction between the objective reality, which is independent of any theory, and the physical concepts with which the theory operates. ... Every element of the physical reality must have a counterpart in the physical theory. ... If, without in any way disturbing a system, we can predict with certainty (i.e. with probability equal to unity) the value of a physical quantity, then there exists an element of physical reality corresponding to this quantity."

To make matters worse, Podolsky also sent a copy to the *New York Times*, implying that quantum mechanics was wrong. Einstein, who had not been entirely satisfied with the way the paper had been written in the first place was very upset by this action and apparently never spoke to Podolsky again.

To some, the arguments presented in this paper, seemed like an escape from the Uncertainty Principle. Alfred Land wrote, "It is unphysical to accept the idea that there are particles possessing definite positions and momenta at any given time, and then to concede that these data can never be confirmed experimentally, as if by a malicious whim of nature."

This paper caused immediate strong reactions and to this day there are philosophers who believe that it is of fundamental importance for the epistemology of quantum physics. The reaction by the physicists already converted to quantum mechanics was also quite strong. Thus, Edward U. Condon found the following objection. "Of course, a great deal of the argument hinges on just what meaning is to be attached to the word 'reality' in connection with physics."

Pauli's reaction was even stronger. "If a student in the early semesters had made such objections to me, I would have regarded him as very intelligent and hopeful." In a letter to Heisenberg sent about a month after appearance of the EPR paper he repeated, "Once again Einstein has expressed his views on quantum mechanics. As we know, each time this happens it is a catastrophe, and this time not in the best of company either." The letter repeats the statement, "Still, if a first-year student had come to me with this question, I would have considered him quite promising."

Immediately after this paper appeared, Rosen visited Kharkov in the Soviet Union. Once there he went to see Lev Davidovich Landau (1908 – 1968) and as he entered that worthy's office, Landau jumped up and shouted, "How could Einstein have written such a stupid thing!"

Léon Rosenfeld later remembered, "This onslaught came down on us as a bolt from the blue. Its effect on Bohr was remarkable. We were then in the midst of groping attempts at exploring the implications of charge and current distributions, which presented us with riddles of a kind we had not met in electrodynamics. A new worry could not come at a less propitious time. Yet, as soon as Bohr had heard ... of Einstein's argument, everything else was abandoned: we had to clear up such a misunderstanding at once." Bohr repeatedly asked, "What can they mean? Do you understand it?"

Six months later, Bohr and his colleagues repelled Einstein's onslaught with a paper of their own. As Bohr then stated, "They do it smartly, but what counts is to do it right."

Although Einstein accepted the force of Bohr's argument, he never relinquished his position that a more complete theory must be possible. In later years he wrote to Max Born. "I cannot provide logical argument for my conviction, but can only call on my little finger as a witness, which cannot claim any authority to be respected outside my own skin." Also, in another letter to Born he repeated, "I can quite well understand why you take me for an obstinate old sinner, but I feel clearly that you do not understand how I came to travel my lonely way. It would certainly amuse you, although it would be impossible."

The Einstein, Podolsky, Rosen (EPR) paper provoked another assault on the idea that quantum mechanics is complete; this time it was "Schrödinger's cat". In a paper in *Die Naturwissenschaften*, 1935, Schrödinger presented what he thought was an argument to show that quantum mechanics is incomplete. This argument so impressed Einstein that in a letter to him in 1939 he wrote, "The prettiest way to show this is by your example with the cat. ....." What Schrödinger argued, after studying the EPR paper, starts with the hypothesis that "states of a macroscopic system which could be told apart by a macroscopic observation are distinct from each other whether observed or not." In other words, if a tree falls in a forest it makes a noise whether or not someone is there to hear it. To "show" that quantum mechanics goes against this he imagined the following "fiendish device".

A cat is enclosed in a steel chamber with some radioactive material that has a 50% chance of decaying in an hour. There is also a Geiger counter in the chamber that, if the material decays, triggers a circuit that kills the cat. Then, according to quantum mechanics, an hour after the cat has been placed inside this chamber, the wavefunction ψ, describing the cat is a fifty-fifty mixture of cat alive and cat dead. Furthermore, if after the hour is up the circuit is disabled, the cat remains in this state of neither alive nor dead until the box is opened and an observation is made. The act of observing then causes the wavefunction to collapse to the observed state of cat alive or cat dead. Since a cat cannot be both alive and dead at the same time, quantum mechanics must be incomplete.

The argument here strikes precisely at the idea that the act of observation is what forces the wavefunction into one of the two possibilities and thus either makes the cat alive or kills it. So, why are most physicists not too concerned with this argument? The reason: they view the firing of the

Geiger counter as a macroscopic observation. It is this triggering (or not) of the Geiger counter that counts as a measurement. So, it is at that instant that the wavefunction for the cat collapses to one of the two possibilities, not at the opening of the box.

Even at the advanced age of 77, Bohr was still trying to explain why our language for quantum physics is inadequate because everything has to be explained in a language adapted to macroscopic physics. He said, "Of course it may happen that when the electronic computers start talking in a few thousand years time their language will be different from ours and they will think us crazy because they will not be able to communicate with us. But that is not our problem."

Later in life Bohr became interested in genetics. He was of the firm belief that in humans hereditary factors are not important. Nevertheless he told the following story on more than one occasion. A father had identical twins, which he raised exactly the same with identical educations until a psychologist told him that this was wrong. They should be allowed to develop their individuality. To allow the boys to develop in completely different fashion the father then sent one off to Harvard where he developed into a perfect Harvard gentleman. The other son he sent to Yale where he developed into a complete Yale cad. Still, there was no difference between the two.

When he was already an elder statesman of physics, Bohr visited the Soviet Union. The Academician Igor Tamm (1895 – 1971) asked him, "How is you have been able to found such a wonderful physics school?" Bohr answered, "We older people have simply not been afraid to appear more stupid in front of them than our younger colleagues."

Regarding the whole development of quantum mechanics, Oppenheimer later recalled, "It was a heroic time. It was not the doing of one man; it involved the collaboration of scores of scientists from many different lands, though from first to last the deep creative and critical spirit of Niels Bohr guided, restrained, deepened, and finally transmuted the enterprise."

Schrödinger had this to say, "As our mental eye penetrates into smaller and smaller distances and shorter and shorter times, we find nature behaving so entirely differently from what we observe in visible and palpable bodies of our surroundings that no model shaped after our large scale experiences can ever be 'true'. A completely satisfactory model of this type is not only practically inaccessible, but not even thinkable. Or to be more precise, we can of course, think it, but however we think it, it is wrong; not perhaps quite as meaningless as a 'triangular circle' but much more so than a 'winged lion'."

He also wrote, "We must not admit the possibility of continuous observation. Observations are to be regarded as discrete, disconnected events. Between there are gaps which, we cannot fill in. There are cases where we should upset everything if we admitted the possibility of continuous observation. That is why it is better to regard a particle not as a permanent entity but as an instantaneous event. Sometimes these events form chains that give the illusion of permanent beings but only in particular circumstances, and only for an extremely short period of time in every single case."

Lev Landau also was very impressed with the subtlety of quantum mechanics, "By studying nature, man can overtake his imagination; he can discover and understand what he is even unable to imagine."

Even so, a small number of physicists continue to this day to be unhappy with the standard (so-called Copenhagen) interpretation of quantum mechanics. There is the so-called Many-worlds Interpretation much favored by cosmologists. Here are some comments by the mathematical physicist Ray F Streater (1936 –) about this interpretation, which he lists among "Lost Causes in Theoretical Physics" on his homepage. "There is nothing to the many-worlds theory. There are no theorems, conjectures, experimental predictions or results of any sort, other than those of Hilbert space. It is not a cogent idea. In fact, it would need a specification of what observables correspond to each operator before anything rich enough to be worth studying arises. It is hoped by its advocates that the existence of so many worlds can be used to justify the law of large numbers, from which one might give the theory a probabilistic interpretation. If this were possible, then we just arrive at quantum mechanics."

There is also the Statistical Interpretation which claims that quantum mechanical statements apply only to ensembles of particles and not to individual particles. This idea may hark back to a (now proven incorrect) statement due to Schrödinger.[3] "We never experiment with just one electron or atom or (small) molecule. In thought experiments we sometimes assume that we do, this invariably entails ridiculous consequences. ... In the first place it is fair to state that we are not experimenting with single particles any more than we can raise Ichthyosauria in the zoo."

As a matter of fact, physicist now routinely work with single electrons. This is especially true in the area of nanophysics.

---

[3]E. Schrödinger, *Br. Jl. for Phil of Sci. 3*, 109, (1952)

Right from the start there were attempts to make unobserved or even unobservable, so-called "hidden variables" responsible for the indeterminacy in quantum mechanics. Von Neumann tried to settle this question with a 'proof' of the nonexistence of hidden variables. However, his proof assumed that the hidden variable were themselves quantum mechanical in nature. Later, John Stewart Bell (1928 – 1990) made this statement in *Omni*, "The proof of von Neumann is not merely false, but foolish!"

Symposium On the Development of THE PHYSICISTS' CONCEPTION OF NATURE at the International Centre for Theoretical Physics, Miramare, Trieste, Italy — Dedicated to P.A.M. Dirac on his seventieth birthday.
Front Row, left to right: P.A.M. Dirac and W. Heisenberg.
Second Row, left to right: John Bell, unidentified person, Josef M. Jauch.

In 1980 John Stewart Bell (1928 – 1990) presented a paper with the title "BERTLMANN'S SOCKS AND THE NATURE OF REALITY"[4] in which he again discussed the EPR paradox. He started by saying, "The

---

[4]Invited address to the meeting of Philosophers and Physicists, Fondation Hugot du Collège de France

philosopher in the street, who has not suffered a course in quantum mechanics, is quite unimpressed by Einstein-Podolsky-Rosen correlations. He can point to many examples of similar correlations in everyday life. The case of Bertleman's socks is often cited. Dr. Bertlman likes to wear socks of different colours. Which colour he will have on a given day is quite unpredictable. But when you see that the first sock is pink you can already be sure that the second sock will not be pink. Observation of the first, and experience of Bertelman, gives immediate information about the second. There is no accounting for tastes, but apart from that there is no mystery here. And is not the EPR business just the same?"

What Bell is getting at is the concept of entanglement. With regard to quantum entanglement Schrödinger said, "[Entanglement] is not one, but rather the characteristic trait of quantum mechanics, the one that enforces its entire departure from classical lines of thought."

What is entanglement? The correlation with respect to Bertlmann's socks is a case of classical entanglement. If the left sock is pink, the right sock is not pink. This is true whether or not we observe the left sock. In quantum mechanics two systems may also be correlated so that one wavefunction describes both, yet neither is in a definite state unlike in the classical case. For example if an electron and positron annihilate they produce two photons that fly apart in opposite directions. If the total angular momentum of the electron-positron pair was zero, then the total angular momentum of the two photons must also be zero. The two photons are entangled. If the photon going to the left has angular momentum pointing up then without having to measure we know that the photon going to the right has angular momentum pointing in the opposite direction—down. However, there is a big difference between this situation and the classical case: Neither photon has a definite angular momentum (up or down) until the angular momentum of one of them is measured. Referring back to Bertlmann's socks it is as if the left sock did not assume the color pink until it was observed. Einstein called this quantum entanglement a "*spukhafte Fernwirkung*" or "spooky action at a distance".

Bell devised a way of testing quantum entanglement and in his famous "Bell's inequality" gave a concrete method (set of experiments) for checking the quantum mechanical prediction. These results gave a way of comparing quantum mechanics against any local theory with an underlying classical reality. He showed that, for certain predictions, any local hidden variable theory had to differ from quantum mechanics. In the early 1980s in a series of beautiful experiments Alain Aspect (1947 –)— while still a graduate student— checked these so-called Bell's inequalities and quantum mechanics passed the test with flying colors. Of course, this has not ended the dispute, even though, after more than eighty-five years of test after test, there is not a single experiment in conflict with quantum mechanics.

The final comment regarding quantum mechanical observations is left to John Archibald Wheeler (1911 – 2008) who stated it in the following concise form. "No elementary phenomenon is a phenomenon until it is a registered phenomenon."

Chapter 15

The Electron Spins

"If I talk in the first person, then there are two reasons. First: lack of modesty, and second: as I tell that history, I can only speak about my own experiences. You know, when Uhlenbeck tells the history of spin then he tells a different story. I don't think either of us lies. But if someone is lying then it is a little more I than he." From S. A. Goudsmit's lecture at the Golden Jubilee of the Dutch Physical Society, April 1971.

The beautiful patterns of the spectra of atoms that led Bohr to his model and eventually Schrödinger to his equation and quantum mechanics still had one more surprise. Actually this goes back to 1896 when Pieter Zeeman (1865 – 1943) at the University of Leyden approached his boss, the famous Heike Kammerlingh Onnes (1853 – 1926), to study the spectral lines of atoms that were placed in a strong magnetic field. He was looking for an effect first studied without success by the great Michael Faraday (1791 – 1867), namely the effect of a magnetic field on light. He believed that if Faraday had thought the attempt worthwhile then it should be done with the better equipment available to him.

Onnes opposed this as a waste of time since Zeeman had been unsuccessful in his earlier attempts. However, Zeeman wanted to use the much bigger electromagnet in Onnes' laboratory. So, when Onnes left for a conference, Zeeman used this magnet (without permission) and found that indeed the lines in sodium broadened when placed in its strong magnetic field. This was the beginning of the Zeeman effect. When Zeeman told Onnes about his wonderful result, this worthy fired him for his efforts. Zeeman was fortunate to then get a position at the University of Amsterdam. Here he was able to show that a magnetic field actually split the sodium lines into two lines of almost the same color. The final sentence in the paper that Zeeman published from Amsterdam to announce this result reads as follows. "I return my best thanks to Prof. Kammerlingh Onnes for the interest he has shown in my work."

Hendrik Antoon Lorentz, interpreted these new lines as being due to a particle whose charge to mass ratio was about 2000 times that of the hydrogen ion. This was, of course the electron, but the interpretation preceded Wiechert's and J. J. Thomson's discovery of the electron by one whole year. This splitting of the spectral lines is now called the Zeeman effect. Zeeman shared the 1902 Nobel Prize in Physics with Lorentz "in recognition of the extraordinary service they rendered by their researches into the influence of magnetism upon radiation phenomena". Later in 1913 Onnes also won the Nobel Prize in Physics "for his investigations on the properties of matter at low temperatures which led, inter alia, to the production of liquid helium".

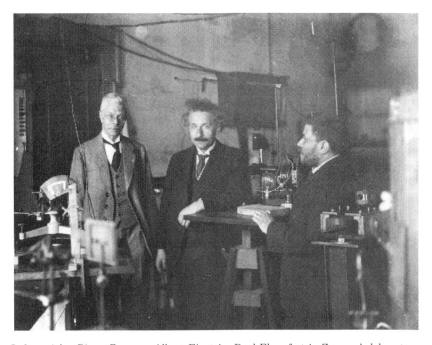

Left to right: Pieter Zeeman, Albert Einstein, Paul Ehrenfest in Zeeman's laboratory.

Faraday, who had inspired Zeeman's efforts, had also discovered electromagnetic induction. When he demonstrated to a lady visitor the tiny current that was produced when he plunged a magnet into a coil of wire, she asked, "What is the use of that?"

He replied, "Madam, what use is a newborn baby?" His "baby" later grew up to become the basis of almost all electric power generated today. Since he had found a relationship between electricity and magnetism Faraday also tried unsuccessfully to establish a connection between electricity and gravity.

When John Tyndall (1820 – 1893) offered the Presidency of the Royal Society to Faraday, he responded, "Tyndall, I must remain plain Michael Faraday to the last; and let me tell you, that if I accepted the honor that the Royal Society desires to confer upon me, I would not answer for the integrity of my intellect for a single year." Near the end of his life he had this to say about his efforts, "Here end my trials for the present. The results are negative. They do not shake my strong feelings of the existence of a relation between gravity and electricity, though they give no proof that a relation exists."

The Zeeman effect remained a mystery for quite a while since only a few lines of the observed splittings could be explained. These constituted the "normal" Zeeman effect. But, according to standard theory, most of these lines observed should not even have occurred and could not be explained at all. These were dubbed the "anomalous" Zeeman effect. The explanation for the anomalous Zeeman effect was published about a quarter of a century later at about the same time as the development of quantum mechanics.

In the period from 1922 to about 1925 there was extreme, but friendly, competition between several physicists—most prominently Alfred Landé (1888 – 1976), Wolfgang Pauli, and Arnold Sommerfeld—to explain the Zeeman effect. Landé gathered as much of the experimental data as were available. After much analysis of these data he constructed a model and was able to obtain a formula that accounted for the number of lines into which a given line would split. However, after more study, he realized that his model did not account for the size of the splitting.

In 1924 Pauli had already rejected the Sommerfeld-Landé model for the multiplicity of spectral lines since it forced Landé to assume that the two electrons in Helium played different roles. In fact he warned Landé, "The very fact that the two electrons in Helium have to play entirely different roles—one the core electron and the other the radiant electron—is the failure of the model." Pauli warned repeatedly against the use of models. He called this a "classic" fault of excessive confidence in a model. He further stated, "I can hardly believe in the models that are currently being considered. Why don't we study the role of multiplet terms just from the experimental results?" Perhaps there was a spiritual link with his godfather, Ernst Mach.

Many years later, Manfred Eigen in the *Physicists Conception of Nature* wrote about models, "A theory has only the alternative of being right or wrong. A model has a third possibility: it may be right, but irrelevant."

Pauli's father was Wolfgang Josef and his mother was Berta Camilla (née Schütz). His father was a medical doctor in Vienna where he converted from Judaism to Roman Catholicism and changed his name to Pauli. He gave up his practice of medicine and, inspired by Ernst Mach, became a university professor in chemistry. Mach became Wolfgang Pauli's godfather.

Young Wolfgang breezed through the gymnasium, even though he paid little attention in class and secretly studied Einstein's papers hidden under his desk. He then studied under Sommerfeld at the Ludwig-Maximilian University of Munich. Shortly after he was joined by Werner Heisenberg who gave the following description of Pauli's lifestyle. "Wolfgang was a typical night owl. He preferred the town, liked to spend evenings in some café, and would thereafter work on his physics with great intensity and great success. To Sommerfeld's dismay he would rarely attend morning lectures and would not turn up until about noon."

Sommerfeld had this to say about Pauli. "I can't teach him anything; at my suggestion he is writing a summary of Einstein's relativity theory." Actually Sommerfeld had been asked to write this article for the *Encyclopädie der mathematischen Wissenschaften* and had passed the work on to Pauli. This book, written by a twenty-one year old, appeared in 1921 and contained the prophetic statement, "Perhaps the theorem on the equivalence of mass and energy can be checked at some future date by observations on the stability of nuclei."

In a review of this book Einstein wrote, "Whoever studies this mature and grandly conceived work might not believe that its author is a twenty-one year old man. One wonders what to admire most, the psychological understanding for the development of ideas, the sureness of mathematical deduction, the profound physical insight, the capacity for lucid, systematical presentation, the knowledge of the literature, the complete treatment of the subject matter, or the sureness of critical appraisal."

There is a story that Einstein and Pauli had met a year earlier when Einstein visited Munich to give a talk. Pauli was still a graduate student at the time. Now, as was the custom, the professors all sat in the front row with successively less important people further away from the front and graduate students at the very back. Pauli seemed unaware of these distinctions of rank and sat in the front row in his Bavarian Lederhosen. At the end of the talk the professors all started clearing their throats, maneuvering for the privilege of asking the first question. Although still only a lowly graduate student, Pauli did not hesitate. He scanned the audience and remarked, "You know, what Herr Einstein has been telling us is not quite as stupid as it might sound."

When Pauli received his doctorate in 1921 Sommerfeld wrote that it demonstrated, "...like his many already published smaller investigations and his larger encyclopedia article, the full command of the tools of mathematical physics."

From Munich Pauli went to Göttingen as Born's assistant. Since Born suffered frequent asthma attacks Pauli was often called upon to give Born's lectures. These were from 11:00 AM to noon. Pauli, however, usually forgot and if Born's maid went to remind him at 10:30 he was still fast asleep since, as already pointed out, it was his habit to work until late at night. Pauli's neighbors in Göttingen worried about him because they saw him every night sitting at his desk rocking like a praying Buddha until the early hours of the morning. Born had to assure the neighbors that he was truly quite normal, just a genius.

Not only Pauli, but many physicists are known for the strange hours they keep. Thus, the response by Robert Sugar at Santa Barbara to the chair's request that he teach an 8:00 AM class was typical, "I don't stay up that late."

Pauli not only loved to stay up late, but also to frequent questionable cabarets. This was another reason he would arrive late for work. It was also the reason Pauli did not like Göttingen with its small town atmosphere and why he left after merely a year there. Born's comment with regard to this was, "He can't stand life in a small town."

In Göttingen Pauli first met Niels Bohr when the latter came for a visit. Of this meeting he said in his Nobel lecture, "...a new phase of my scientific life began when I met Niels Bohr personally for the first time. This was in 1922, when he gave a series of guest lectures at Göttingen when he reported on his theoretical investigations on the periodic system of elements. During these meetings, Bohr asked me whether I could come to Copenhagen for a year." Pauli accepted and spent the year 1922-23 in Copenhagen. This is where, in his own words, he "made a serious effort to explain the so-called 'anomalous Zeeman effect', ...a type of splitting of the spectral lines in a magnetic field which is different from the normal triplet."

At the beginning of 1924 Pauli wrote in a letter to Bohr regarding the use of half-integer quantum numbers. "The atomic physicists in Germany today fall into two groups. The one calculates a given problem first with half-integer values of the quantum numbers, and if it doesn't agree with experiment they then do it with integral quantum numbers. The others calculate first with whole numbers and if it doesn't agree then they calculate with halves. But both groups of atomic physicists have the property in common that their theories offer no a priori reasoning which quantum numbers and which atoms should be calculated with half-integral values of the quantum numbers and which should be calculated with integral values. I myself have no taste for this sort of theoretical physics, and retire from it to my heat conduction of solid bodies."

In his studies of the anomalous Zeeman effect Pauli had come up against a problem that required a language that had not even been invented. His frustrations were enormous. At this time he was in Copenhagen at Niels Bohr's Institute and to clear his mind he would take long walks in the evening. His frustration must have been evident for, as he later recalled, one evening as he plodded beside the water he heard an elderly lady call to him, "Don't jump in, young man. No woman is worth committing suicide over."

Many years later in a talk at Princeton he recalled that on another occasion he met a friend as he was again pondering this problem while wandering the streets. His friend commented to him that he looked very unhappy. Pauli's response was, "How can I not look unhappy; I don't understand the Zeeman effect." It was also about this time (1925) that Pauli made his famous statement, "Physics is at the moment once again very wrong. For me, in any case, it is much too difficult and I wish I were a film comedian or something similar and had never heard of physics." Of course, this was when Charlie Chaplin was all the rage and it was also very different from the cocky young Pauli who when he first arrived in Copenhagen had proclaimed, "I won't have trouble with physics, but what I'm really afraid of is the Danish language."

At this time Pauli also became depressed by the lack of understanding of quantum phenomena and in a letter to his friend Heisenberg expressed the wish that he had become a cobbler or a tailor. A few months later, after Heisenberg had made some start into matrix mechanics and sent these results to Pauli, the latter was again delighted that he had studied physics and said that he saw, "... some light at the end of the tunnel."

Soon after quantum mechanics raised his spirits, Pauli had two terrible years. Always close to his mother, her suicide in 1927 was a great tragedy to him. Even worse, his father remarried soon after. In a clear reference to the Brothers Grimm fairy tales, Pauli referred to his father's new wife as "the evil step-mother". His happiness should have resumed when he married Käthe Margaret Deppner just before Christmas in 1929. The marriage unfortunately was a failure from the start and less than a year later they divorced. After the divorce Käthe married the chemist, Paul Goldfinger, whom she had known even before her marriage to Pauli and with whom she had spent a lot of time in Berlin, even during her marriage. Pauli was upset by this marriage and said, "Had she taken a bullfighter I would have understood, but an ordinary chemist . . . ."

Pauli went on to become Professor of theoretical physics at the Eidgenösische Technische Hochschule in Zurich. His scientific successes were many, but around 1933 his personal problems increased and he started to drink heavily. One story has it that at dinner with a couple he ordered two bottles of wine and when the wine arrived declared, "These are for me." To the couple's surprise he then proceeded to drink them by himself.

At about this time he entered analysis under Sigmund Freud's rival Carl Gustav Jung. Pauli related details of more than 1000 dreams to Jung who eventually published a book based on some 400 of these dreams. Pauli's life improved after he married Franciska Bertram in 1934.

Again fate intervened. After Germany annexed Austria, and although living in Switzerland, Pauli found himself a German citizen. This was rather awkward after war broke out in 1939. When Princeton University offered him a chair in Theoretical Physics he gladly accepted. After the war he returned to Zürich. In 1945 he received the Nobel Prize in Physics "for the discovery of the Exclusion Principle, also called the Pauli Principle".

Pauli's study of multiplet terms in the spectra, just from the experimental results, led him to conjecture and to publish that the origin of the multiplicity lay in each electron by itself; that each electron had a "classically indescribable two-valuedness". This idea fit in well with the Pauli exclusion principle and helped to explain why some lines were missing in the observed spectra. To verify his ideas, Pauli went to see Landé to tap his vast amounts of experimental data. There he met a young man, Ralph de Laer Kronig (1904–1995) who had the bright idea that this two-valuedness might be due to an electron rotating about itself. This model agreed with Pauli's calculations and the experimental facts, except for a factor of two in the intensity. Pauli was very cool to this idea and so were Bohr and the people at Copenhagen when Kronig went there to talk about his ideas.

Since there was still this discrepancy of a factor of two, as well as the fact that if one made a model of a rotating spherical electron the surface of this electron would have to move at 10 times the speed of light to give the right angular momentum, both Bohr and Pauli rejected Kronig's model. As a consequence Kronig never published his results.

Pauli was indeed a harsh critic of his own work as well as of that of others. He even once commented to an assistant, "I don't mind if you think slowly Herr Doktor, but I do object if you publish faster than you think." On another occasion he rejected a paper for publication with the words, "This is not right. It is not even wrong." His assistant, Victor Weisskopf had this to say, "It was absolutely marvelous working for Pauli. You could ask him anything. There was no worry that he would think a particular question was stupid, since he thought all questions were stupid."

Another time, many years later, after Emilio Segré (1905 – 1989) had presented a lecture on proton-proton scattering, Pauli left the lecture hall with him and another physicist. Pauli turned to Segré and told him, "I have never heard a worse speech than yours." After a moment's reflection he turned to the other colleague and continued, "Except when I listened to your inaugural speech in Zurich."

Hendrik Bugt Casimir was Pauli's assistant in 1932-33. On a car trip with Pauli driving, Casimir feared for his life and made some disparaging remark. Pauli then made a deal with him. "I won't criticize your physics if you won't criticize my driving." Many years later when Casimir was director at Philips and met with Pauli again, he asked, "How is your car?" Pauli rejoined, "I gave up driving and you gave up physics."

There is a slightly different version of this story as told by Casimir in his autobiography, *Haphazard Reality*. After Casimir had decided to remain at the Philips Research Laboratory, Pauli would advise any physicist who was about to visit Casimir to address him as "Herr Direktor" since this "annoys him terribly". In 1958 Pauli visited Philips for the last time. As Casimir, Pauli and some friends were sitting together reminiscing, someone asked Casimir if he'd had a tough time as Pauli's assistant. At a loss of what to say, Casimir reported that after Pauli got his driver's license they agreed that if he would refrain from criticizing Pauli's driving then Pauli would not criticize his physics. He continued, "Now I do not want to brag about my physics, but I do believe that in those days it was slightly better than Pauli's driving." After the laughter, Pauli agreed, "Yes, it may have been like that. But I don't drive anymore. And you, Herr Direktor, aren't doing physics anymore. *Die Sache stimmt noch immer.*" (Things are still in accord.)

Wander Johannes de Haas (1878 – 1960) also warned Casimir, "You have convictions? Dangerous thing for an experimenter to have. Better get rid of them." de Haas is remembered for the Einstein-de Haas Effect (1916) in which a small iron rod suspended by a thin quartz fiber is brought into oscillation by a variable magnetic field in the same direction as the axis of the rod. According to Casimir there is some doubt as to whether this effect was observed by de Haas at that time, but he was still instrumental in bringing this effect to the attention of the physics world.

When he was close to eighty de Haas confessed to Casimir. "I have always been a heathen, but now I don't have much longer to live and I have instructed Woltjer to convert me. But he is not making any progress. He reads from the Bible beautifully but he is not making any progress. He should hurry up."

Getting back to Pauli, a colleague, whose work Pauli had criticized, had given him directions to some place and asked him the following day whether he had found it. "Oh yes. You express yourself quite intelligibly when you don't talk about physics."

Pauli had the disconcerting habit that during a seminar he would sit in the front row and wag his head from side to side as if he were negating everything said. For those who knew him this was a sign that Pauli agreed with what he heard. It was only when he stopped shaking his head that he was about to loosen one of his devastating questions. Gunnar Källén, who knew Pauli well, took advantage of this to anticipate Pauli. While speaking, he kept his eyes on the latter and as soon as Pauli stopped shaking his head would exclaim, "Well, here is another point that Pauli has not understood."

Although the consummate theorist, Pauli was very aware of the experimental aspects of physics. In a letter to Eddington dated, Sept. 20, 1923 he made the following assertion. "It has often been said that one could retain the Maxwell equations in the vacuum (charge-free space). Only the interaction between light and material systems could not be treated with the help of the classical theory. In fact the two are inseparable. Only by means of this interaction can the electromagnetic field of a lightwave be defined. The Maxwell equations in a vacuum are neither correct nor incorrect; they are meaningless in every application where the classical theory fails and where it is not a question of statistical averages."

In his later years Pauli became more of a mathematical physicist and cautioned, "Never believe what experimentalists say first." He repeated, "Never work too closely with experimenters. Allow the results to settle."

Another phenomenon for which Pauli was well known and of which he was almost as proud as his Exclusion Principle was the Pauli Effect. The Pauli Effect is the phenomenon that no experiment could succeed if Pauli was anywhere in the vicinity. For example, in Professor James Franck's laboratory in Göttingen a complicated piece of apparatus for studying atomic phenomena collapsed one day without reason. In a humorous vein, Franck (1882 – 1964) then wrote to Pauli about this and asked him if he had anything to do with it. Pauli wrote back that the train on which he was traveling on his way to Copenhagen had stopped at that precise moment in the railroad station in Göttingen.

An extension of the Pauli Effect may also have occurred on the following occasion. At one of the Solvay conferences when Pauli was still quite young and not yet famous he was asked to give his talk, which had been scheduled for the afternoon, in the morning instead. Pauli, who always slept until almost noon, declined. He was informed that the other speaker was more important and that if he was not willing to change times then his talk would have to be cancelled. Pauli still refused and his talk was cancelled. At the time originally scheduled for his talk, Pauli walked into the lecture theatre just as the chairman, who had cancelled Pauli's talk, finished introducing the speaker and sat down. The chair collapsed under the chairman and Pauli was heard to laugh diabolically.

The Hamburg astronomers once invited Pauli to visit their observatory. Aware of the effect named after him he declined with the words, "Telescopes are very expensive." The astronomers assured him that the Pauli Effect was not operative in their observatory. When Pauli entered the dome there was a large crash. After all of them had recovered their equanimity they discovered that the large cast iron cover had fallen off one of the telescopes and shattered on the concrete floor.

Another good example of the Pauli effect occurred at a reception where there was to be a staged Pauli effect. A chandelier was suspended by a rope and the chandelier was to be allowed to crash to the floor after Pauli entered the room. As it happened the rope going over the pulley got wedged and nothing happened.

While in Copenhagen Pauli, who always tended to be overweight, was gaining even more weight. As a consequence Dirac, who was deemed to have the most self-discipline, was asked to watch Pauli to make sure he did not overindulge. In jest Pauli asked, "How many lumps of sugar may I put in my coffee?" Dirac replied, "I think one is enough for you," after a pause, "I think one is enough for anybody," after a further pause, "I think lumps of sugar are made in such a way that one is enough for anybody."

On another occasion after a much slimmed down Pauli returned to Copenhagen everyone noticed that he was much grumpier and, unlike his usual self, without a sense of humor. This was remedied by taking him to the local pub for sausages and beer. In a week Pauli was back to his old self.

At the Leyden Observatory, September 1923.
Front row: Sir Arthur Eddington and Hendrik Antoon Lorentz.
Back row: Albert Einstein, Paul Ehrenfest and Willem de Sitter.

Carlo Bernardini attributes his good relations with Pauli to the following. "I had the honor and pleasure to be almost a friend of Pauli, mainly because he liked wine and I went with him near Frascati, and then around Florence, to drink good wine."

After Hendrik Antoon Lorentz retired at the University of Leyden his chair was offered to Ehrenfest who turned it down because he did not consider himself good enough. Finally, however, Lorentz himself convinced Ehrenfest to accept the position and he turned out to be an excellent teacher and researcher. He used to tell his students that "physics is simple but subtle". Although Ehrenfest stated that he never worked on the "grand problems" of physics he did clear up many important points.

Once settled in Leyden, Ehrenfest learned to speak Dutch, but seemed never able to properly pronounce "viertien"—the number 14. He always pronounced it "vier teen"—four toes. His students would laugh, and he never knew why. (The vowel sound in the syllables vier and tien is close to that in the English word "see", whereas in the syllable teen the vowel is close to the English wood "say".)

According to Casimir, Ehrenfest was a "merciless critic of the stupid and unclear". He could also, like Pauli, be biting in his sarcasm. Thus, when Ernst Rupp gave a talk with the somewhat bombastic title "An electric analog of the Compton effect" at the *Deutsche Physikalische Gesellschaft* on some experiments with colliding beams, Ehrenfest made the following comment. "In the same way, when I shoot off a bird's tail I might call it a biological analog of the photoelectric effect." Rupp had also as early as 1928 published two papers in which he presented pictures as experimental proof of electron diffraction. Ten years later careful work by physicists at the University of Munich showed that these pictures had to be fraudulent since the results could not be duplicated.

When Ehrenfest and Pauli met for the first time (probably at the 1927 Solvay conference), Ehrenfest was immediately repelled by Pauli's arrogance and abrupt manner. This led him to state, "I very much admire all that you have written but not what you are." Pauli's response was, "Strange, my reaction to you was just the reverse." Later Ehrenfest dubbed Pauli, *die Geissel Gottes* (the scourge of God), a reference to Attila the Hun. Pauli apparently was very proud of this appellation.

Hendrik Casimir in his biography, *Haphazard Reality*, relates the following contest of wits between Pauli and Ehrenfest. On October 31, 1931, Pauli was awarded the Lorentz medal of the Royal Netherlands Academy of Arts and Sciences. On this occasion Ehrenfest, who normally did not worry about dress conventions, had written to Pauli that he should wear a black suit for this affair. Pauli replied on a postcard.

Zürich, October 26, 1931.

Dear Ehrenfest,

Just now I ordered a new black suit at my tailor's. But I shall only put it on in Amsterdam, if you promise me to thank me in public, in your official allocution at the Academy, for not having saved myself the trouble of going to the tailor. ....

Ehrenfest kept his promise in the following fashion. After discussing the Pauli Exclusion Principle and how it keeps electrons in an atom apart so that atoms are much bigger than they would be without the exclusion principle he finished by saying, "What then prevents the atom from making itself smaller in that way? Answer: Only the Pauli principle: 'No two electrons in the same quantum state.' Therefore, the atoms so unnecessarily fat: Therefore the stone the piece of metal, etc. so voluminous. You will have to concede Herr Pauli, by partially waiving your exclusion principle you might free us from many worries of daily life, for instance from the traffic problems in our streets and you might considerably reduce the expenditure for a beautiful, new, formal black suit."

Paul Ehrenfest's family came from a small village in Moravia, but moved to Vienna where Paul was the youngest of five children. His four older brothers encouraged and supported him so that by the time he was ready to enter school he could already read and write. When he was ten his mother died of breast cancer and his father remarried to his first wife's younger sister, Josephine Jellinek who was about the same age as Paul's oldest brother, Arthur.

The gymnasium was not a happy place for young Paul; the only subject in which he did well was mathematics. To further complicate his life, his father died when Paul was sixteen. Arthur became his guardian and persuaded Paul to stay in school. At the time Paul was depressed to the point of contemplating suicide. He graduated from the gymnasium in 1899, but his experiences left him scarred and later in life he insisted on home schooling for his children. That same year he entered the Technische Hochschule in Vienna where he listened to Boltzmann lecture on the mechanical theory of heat. These lectures inspired an enduring love of mathematics and physics in him.

In 1901 he followed the path of many famous physicists and moved to Göttingen where he studied under the greats: Klein, Hilbert, Stark, Nernst, Schwarzschild, and Zermelo. This is also where he met his future wife. Paul noticed a young Russian student, Tatiana Alexeyevna Afnassye who also attended the lectures of Klein and Hilbert. When she failed to show up at the meetings of the mathematics club he asked about this and was told that women were not allowed to attend. Ehrenfest set out to change this and eventually succeeded. This was the beginning of their friendship. Eighteen months later, in 1904, Paul returned to Vienna and obtained his doctorate under Boltzmann. Tatiana joined him there that same year and they got married. Since Tatiana was Russian Orthodox and

Paul was Jewish they could not legally marry. Neither wanted to adopt the other's religion, so they declared themselves to be without religion in order to be allowed to legally live together. This had serious consequences because a person without religious affiliation could not be appointed at any university in the Austro-Hungarian Empire.

Without a position, Ehrenfest and his wife returned to Göttingen in 1906 with the hope of getting some appointment. While there, Ehrenfest learned of the tragic suicide of his beloved teacher and assumed the task of writing Boltzmann's obituary.

When Einstein left his position in Prague he tried to get Ehrenfest appointed as his successor. Einstein urged Ehrenfest to declare his religious affiliation, but Ehrenfest refused and so lost that possibility.

Next they spent five years in St. Peterburg, still without a position. During this time Paul and Tatiana completed a review article on statistical mechanics for the *Encyclopedia of Mathematics*. Felix Klein had suggested this article to them while they were still in Göttingen. In the meantime Ehrenfest also undertook a tour of German universities to see if he could land something. Fortune finally smiled on him.

Hendrik Antoon Lorentz was about to retire from his position at Leiden and a search for his successor was on. Sommerfeld recommended Paul Ehrenfest. "He lectures like a master. I have hardly ever heard a man speak with such fascination and brilliance. Significant phrases, witty points and dialectic are all at his disposal in an extraordinary manner ... He knows how to make the most difficult things concrete and intuitively clear. Mathematical arguments are translated by him into easily comprehensible pictures." When the telegram announcing that he was appointed as professor at Leiden reached Ehrenfest in St. Petersburg, he had doubts about his abilities to succeed Lorentz. However, Lorentz himself finally persuaded him to accept and Ehrenfest remained in Leiden until his death.

All his life Ehrenfest suffered from feelings of inadequacy, even though he was greatly esteemed by his colleagues. At the beginning of the 1930s his feelings of inadequacy increased. The rise of the Nazis must also have been very disturbing to the man who as a youth had experienced anti-Semitism first hand. His state of mind is evident from a letter he wrote to Bohr in 1931. "I have completely lost contact with theoretical physics. I cannot read anything any more and feel myself incompetent to have even the most modest grasp about what makes sense in the flood of articles and books. Perhaps I cannot at all be helped any more."

A further source of sadness for Ehrenfest was his son Vassily (Wassik) who had been born with Down's syndrome. Wassik not only had severe mental problems, but also severe physical problems. On a visit to the Professor Watering Institute in Amsterdam where Wassik was being treated, Ehrenfest shot his son and himself in the waiting room. Prior to this, Ehrenfest had written two very sad letters: one addressed to his "dear friends: Bohr, Einstein, Franck, Herglotz, Joffé, Kohnstamm, and Tolman!" but never sent. In this letter he ended by saying, "I have no other practical possibility than suicide, and that after having first killed Wassik. Forgive me ... May you and those dear to you stay well."

He wrote a similar letter to his students, but also never sent it.

His friends and colleagues mourned Ehrenfest's death deeply. His humor and kindness had endeared him to all. Although Ehrenfest stated that he never worked on the "grand problems" of physics he did clear up many important points. This is what Einstein had to say about him. "He was not merely the best teacher in our profession whom I have ever known; he was also passionately preoccupied with the development and destiny of men, especially his students. To understand others, to gain their friendship and trust, to aid anyone embroiled in outer or inner struggles, to encourage youthful talent—all this was his real element, almost more than his immersion in scientific problems."

Samuel A. Goudsmit (1902 – 1978) was a student with Ehrenfest. That he was destined for a life in physics was clear from the start since, as a young man, he lived in house with the number 137 [1]. Since Goudsmit was more inclined towards experimental than theoretical work Ehrenfest sent him to work with Friedrich Paschen (1865 – 1947) in Bonn who later became head of the Physikalisch Technische Reichsanstalt, the PTR (sort of like the Bureau of Standards in the USA). The hydrogen spectrum series starting with $n = 3$ is called the "Paschen series" like the one with $n = 2$ is called the "Balmer series".

When he was the head of the PTR one of Paschen's scientists came one day to ask for permission to put his beehives on the flat roof of the building. After a little thought Paschen declined to give permission with the explanation that, "It would set a precedent and soon someone would want to keep chickens and then sheep would follow and the next thing you know there would be cows on the roof of the PTR."

While at Paschen's institute Goudsmit saw more of the experimental results pertaining to the Zeeman effect and came up with a model similar

---

[1] 137 is the reciprocal of the Sommerfeld's fine structure constant, which controls all of electrodynamics.

to that of Kronig. It was based on coupling the internal angular momentum with the fourth quantum number suggested by Pauli. As a consequence he could show that some of the lines in the Zeeman spectrum that were forbidden according to the Sommerfeld-Landé model, but were in fact observed, were not only allowed, but were to be expected. However, he did not have the theoretical abilities to work out all the consequences and sent a note to Copenhagen to Kramers and Kronig for their opinions. The only answer he received was from Kronig who totally ignored what Goudsmit had written. Perhaps he did not want to hurt his feelings about a model that he himself had conceived and rejected due to Pauli's criticism.

George Uhlenbeck (1900 – 1988), another student of Ehrenfest, was a good theorist, well versed in classical physics. So Ehrenfest put him and Goudsmit together to learn from each other. It was Pauli's idea of "classically indescribable two-valuedness" that led them to come up with the model of the spinning electron, the concept initially so opposed by Pauli. Unfortunately their result, just like Kronig's was out by a factor of two. Pauli immediately damned their work as being wrong and they hesitated to publish it. Ehrenfest suggested that they go talk to his predecessor, Hendrik Antoon Lorentz. This great man told them that it was quite impossible. He had looked at such a model some time ago and discovered that the surface of the electron would have to move at ten times the speed of light. As a consequence they returned to Ehrenfest to tell him that they wanted to withdraw their paper from publication. Ehrenfest told them that their paper was either nonsense or very important and that at any rate he had already sent it off to be published. "You both are young enough without a reputation to lose and can permit yourself a folly." Thus, their paper was published in *Naturwissenschaften*.

After Goudsmit and Uhlenbeck published a second paper in *Nature*, Kronig sent a note to this journal. In this note he criticized the shortcomings of the work and finished with the sentence, "The new hypothesis, therefore, appears to effect the removal of the family ghost from the basement to the sub-basement instead of expelling it definitely from the house."

Even Heisenberg, who was usually extremely polite, wrote a letter to Goudsmit and Uhlenbeck and congratulated them on such a "courageous" paper and asked them how they got rid of the factor of two. Clearly, the paper was not taken as being very serious. Uhlenbeck seems to have had misgivings, but Goudsmit did not.

While Pauli was still damning this whole business, a paper by Llewellyn Hilleth Thomas (1903 – 1992) appeared and cleared up this mysterious

Niels Bohr and Wolfgang Pauli studying the tippe top at the inauguration of the new
Institute of Physics at Lund, Sweden.

factor of two. What Goudsmit and Uhlenbeck had failed to realize was
that the electron rotating about the nucleus was moving at relativistic
speeds and that one has to correct for this by using relativity. Thomas had
done the calculation and found that the result gave a correction of exactly
a factor of two.

Soon after this Goudsmit received a postcard from Pauli stating, "I
now believe in the idea of the self-rotating electron." The idea had received
Pauli's imprimatur. Receiving his approval was known as receiving "Pauli's
sanction" since, as already stated, he was a very severe critic of his own as
well as of everyone else's work. Some mathematical niceties, as to how best
to describe this spin, still had to be resolved and this was done by Pauli. As
a consequence the matrices which now describe spin in the nonrelativistic
Schrödinger equation are called the Pauli spin matrices.

In a letter to Goudsmit, Thomas wrote, "I think you and Uhlenbeck
have been very lucky to get your spinning electron published and talked
about before Pauli heard of it. ... More than a year ago Kronig believed

in the spinning electron and worked out something; the first person he showed it to was Pauli—Pauli ridiculed the whole thing so much that the first person became also the last."

A few years later Goudsmit heard Arthur Eddington's lecture on the fine structure constant and asked his older colleague H. A. Kramers if all physicists went off the deep end as they grew older. Kramers reassured him, "No Sam, you don't have to be scared. A genius like Eddington may go nuts, but a fellow like you just gets dumber and dumber."

Eddington's later eccentric ideas were well-known. He once started a lecture, in 1938, with the statement, "I believe that there are 15 747 724 136 275 002 577 605 653 961 181 468 044 717 914 527 116 709 366 231 425 076 185 631 013 296 protons in the universe and the same number of electrons." He also claimed that, "Fewer than a thousand people can understand Einstein's theory of relativity and fewer than a hundred people can discuss it intelligently."

In an address to the American Physical Society in 1976 Goudsmit stated, "It was a little over fifty years ago that George Uhlenbeck and I introduced the concept of spin. It is therefore not surprising that most young physicists do not know that spin had to be introduced. They think that it was revealed in Genesis or perhaps postulated by Sir Isaac Newton, which most young physicists consider to be about simultaneous."

Victor Weisskopf and Dirac also had a discussion about Pauli's opposition to the introduction of the concept of spin.

Dirac: "Pauli very often bet on the wrong horse when a new idea was introduced. He was very much opposed to the positron idea to begin with, for about three months, or was it six months?"

Weisskopf: "Yes. He was also opposed as I said, to your idea of filling the vacuum."

When Pauli visited Hamburg, his reputation as a critic led Wilhelm Lenz (1888 – 1957) to introduce him as follows, "Pauli has often called something wrong that turned out to be right, but he has never judged anything to be right that later turned out to be wrong."

Here are some more Pauli stories and comments.

In talking to young people who were not paying attention Pauli admonished, "I have more experience than you. I have once been young, but you have never yet been old."

His response to Lev Landau, after a long argument when this gentleman pleaded for an admission that not everything he had said was complete nonsense, Pauli had this to say. "Oh no, far from it. What you said was so confused one could not tell whether it was nonsense or not."

During a lecture a student complained to Pauli that he had told them that a certain result was trivial but, the student said he could not understand it. Pauli left the lecture room, returned several minutes later and said, "It is trivial!"

Paul Scherrer (1890 – 1969) proudly showed Pauli how in his lecture he had found a simple explanation for some phenomenon. "You see, here the spin is up and there it is down, and then they interact. Isn't that simple?" Pauli responded, "Simple it is, but it is also wrong."

When Weisskopf showed Pauli a paper on a subject of Pauli's interest that had just been published, the latter said, "Yes, I thought of that too, but I am glad that he worked it out, so that I don't have to."

During all the time that Walter Elsasser (1904 – 1991) studied quantum mechanics with Pauli, the latter ignored him. Five years later Pauli explained to him, "You seemed so weak and shaken up at the time, I was worried that you would faint and keel over if I breathed hard on you."

As reported by Max Dresden (1918 – 1997) in 1988 at a conference in Logan Utah, Pauli summarized the results of World War Two as follows. "The only important thing that happened is that Onsager solved the two-dimensional Ising model."

When Pauli was about to die and was brought to the hospital he asked for the number of his room. When he was informed that it was 137 [2] he responded, "I shall not leave this room alive." His prediction proved correct.

When Schrödinger heard of Pauli's death he mused to his assistant Leopold Halpern, "I don't understand why he was never elected to membership in the Austrian Academy— I wonder if the reason might perhaps be sought in anti-Semitism?" Halpern then pointed out that Pauli had been very open in his criticism of Austrian physicists. After several vigorous puffs on his pipe Schrödinger conceded, "Pauli was a very honest human."

The final comment about Pauli is from one of his erstwhile teachers, Max Born. "Since the time when he was my assistant in Göttingen, I knew he was a genius, comparable only to Einstein himself. As a scientist he was, perhaps, even greater than Einstein. But he was a completely different type of man who, in my eyes, did not attain Einstein's greatness."

---

[2] As already stated earlier $1/137$ is the fine structure constant.

Chapter 16

Epilogue

"Brief is this existence, as a fleeting visit in a strange house. The path to be pursued is poorly lit by a flickering consciousness, the center of which is the limiting and separating I." Albert Einstein 1954.

Spin was the final puzzle in non-relativistic quantum mechanics. In 1928 Dirac finally explained its origin when he created a relativistic wave equation—the Dirac equation—to replace the Schrödinger equation. This equation not only predicted the spin and correct magnetic moment of the electron, but also predicted the existence of a totally new particle, the positive electron or positron. The Dirac equation opened a whole new era of physics and led to what is now called quantum field theory.

There was one other property of spin that remained a mystery for some time. Somehow spin was connected with the Pauli Exclusion Principle. Pauli explained this in 1940, when he showed that relativistic quantum field theory requires a deep connection between the spin of a particle and the symmetry of the wave function for a collection of identical particles. He found that the wavefunction for identical particles whose spin is 1/2 times Planck's constant[1] is antisymmetric if two of these particle are interchanged. In other words, an interchange in the wavefunction of the labels of any two particles makes the wavefunction the negative of what it was. These spin 1/2 particles are called fermions and satisfy the so-called Fermi-Dirac statistics which take this antisymmetry into account. Pauli considered relativity theory essential for this proof and wrote, "In conclusion we wish to state, that according to our opinion the connection between spin and statistics is one of the most important applications of the special

---

[1]actually 1/2 times Planck's constant h divided by 2π

223

relativity theory."

The Pauli Exclusion Principle, used to explain the periodic table, is a direct consequence of this antisymmetry. On the other hand, for identical particles of integer spin the wave function is symmetric under such an interchange. These are called bosons and satisfy the Bose-Einstein statistics.

The origin of these statistics is due to Satyendra Nath Bose (1894 – 1974). In the early 1920s Bose, a professor at the University of Dhaka, gave a new derivation of Planck's blackbody radiation formula by assuming one could have any number of photons with the same energy—in the same state, but could not get his paper published in any of the European journals. He finally sent his manuscript to Einstein asking for his advice. Einstein saw the significance of this work and translated it from English into German and got it published in 1924 in *Zeitshrift für Physik*, one of the most prestigious journals of the time. This was two years before quantum mechanics was developed and before the notion of a wavefunction even existed. Later, in 1926, Einstein extended the notion of what Bose had done from photons (massless particle) to massive particles.

The idea of identical particles is much more subtle in quantum mechanics than in classical mechanics. Consider the red balls in a game of snooker. We think of them as being identical. Yet, when the cue ball scatters them, we can still follow the trajectory of each individual red ball even though they are identical. They are identical, but nevertheless distinguishable. In quantum mechanics, identical billiard balls are totally indistinguishable; one wavefunction describes all the balls. We cannot follow the trajectory of any given ball, only the evolution of this wavefunction as a whole. This prevents us from saying that this red ball was here and that one was there. The wavefunction is the same if we exchange any two identical balls. So, we cannot know which ball is which. If while we looked away, someone interchanged two identical balls we would not be able to tell. In quantum mechanics we cannot tell which ball is which even if we watch the balls—in this case the wavefunction— all the time. In quantum mechanics, identical particles are truly indistinguishable. Even if you were to "watch" the wavefunction all the time you would not be able to follow a particular ball.

The connection between the spin of a particle and the corresponding symmetry of the wavefunction together with the introduction of spin into the Schrödinger equation completed the last phase of non-relativistic quantum mechanics. Now all microscopic phenomena, at energies such that relativity could be neglected, were ready to be explored and exploited. All the forces involved in chemical processes, all the forces involved in the study of materials whether inert or living, all the forces affecting daily life, were in principle known. It was as we already quoted Dirac, ". . . indeed a time when second rate physicists could do first rate work."

This time, unlike at the turn of the century, physicists had no opportunity to claim that "all is understood." Even before they completed the development of quantum mechanics new problems confronted them. There were strong difficulties that arose from trying to understand the meaning of the Dirac equation.

There was also the puzzle, already found by Rutherford, of the extraordinarily strong force that held the atomic nucleus together. This force—called the "strong force"—has continued to be a problem even until today. There are fantastically good models that allow accurate predictions, but these models still lack complete vindication. In this regard it may be wise to again recall the words of Manfred Eigen (1927 –) in the *Physicists Conception of Nature*, "A theory has only the alternative of being right or wrong. A model has a third possibility: it may be right, but irrelevant."

At present the giant CERN accelerator in Geneva is being used to try to verify this model of the strong force by looking for an essential ingredient, a particle known as the "Higgs" particle. If found, this would go a long way to raise this model to the status of a theory.[2]

Another problem confronting physicists of the 1930s was that the nucleus was unstable; it emitted three kinds of radiation: alpha, beta and gamma. George Gamow (1904 – 1968) was able to explain alpha radiation by using tunneling, a direct result of ordinary quantum mechanics. Particularly difficult to understand was beta decay—the emission of electrons. Not only was it impossible to understand how electrons could exist within a nucleus, but even worse, energy as well as momentum conservation appeared to be violated during beta decay. Gamma radiation involves the emission of very hard photons and seemed possible to understand as the transition between different energy levels in a nucleus.

The problem of beta decay remained unsolved until 1967 when Sheldon Lee Glashow (1932 –), Abdus Salam (1926 – 1996), and Steven Weinberg ((1933 –) independently succeeded in unifying electromagnetic and weak forces to produce the "electroweak" theory. This unification is an achievement comparable to Maxwell's unification of electric and magnetic forces to produce modern electromagnetic theory.

Another problem that remained was, how to unite, in a consistent man-

---

[2]Some of these more recent theories are discussed in my book, *From Quanta to Quarks: More Anecdotal History of Physics*.

ner, the theory of special relativity with quantum mechanics? Although elegant solutions (in the form of quantum field theory) were suggested, they seemed to create more problems rather than solve them.

Along these lines are extremely ambitious attempts to unify all forces and gravitation. The most ambitious of these, but so far unsuccessful in terms of experimentally verifiable results, is "string theory". In this theory, particles are viewed as specific vibrations of subatomic strings or membranes. Feynman dismissed it as "crazy, nonsense," and "the wrong direction" for physics. Sheldon Glashow compared string theory to a "new version of medieval theology," and campaigned to keep string theorists out of his own department at Harvard. When he failed, he left. So far, string theory has produced many very elegant mathematical results, but lacks contact with experimental results.

The biggest impact of physics on everyday life came, however, from applying non-relativistic quantum mechanics to systems of particles—to the study of condensed matter. Today we accept devices such as lasers, microchips, cell phones, etc. as commonplace; they are the result of combining non-relativistic quantum mechanics with the imagination of experimentalists, technicians, and engineers. The new problems have yielded rich areas of investigation.

At present quantum mechanics is also yielding extremely interesting results and providing even more interesting question when applied to problems in biology.

It seems that the more we understand, the more there is to understand. Once again physicists have found that one never *solves* problems; one only *creates* new problems. Some time ago with reference to string theory, Steven Weinberg wrote, "How strange it would be if the final theory were to be discovered in our own lifetimes!" This, it seems, is unlikely to happen.

Rutherford would have been pleased "that the enterprise looks like going on forever."

Chapter 17

Glossary and Timeline

17.1 Glossary

Alpha particles (α-particles). Also known as alpha rays, α-particles are the nuclei of helium atoms.

Avogadro's hypothesis states that equal volumes of different gases at the same temperature and pressure contain the same number of molecules.

Avogadro's number is the number of molecules of gas in 22.411 liters at $0°C$ and 1 atmosphere of pressure and is equal to 6.02×10^{23}.

Balmer series is the series of spectral lines that are in the visible region of the spectrum of hydrogen.

Beta particles (β-particles) are electrons. They are also known as beta rays.

Blackbody radiator is an idealized perfect radiator that absorbs all radiation falling on it. The name derives from the fact that any black non-reflecting surface is a good approximation of such a radiator.

Blackbody radiation is the radiation emitted from a blackbody radiator held at a constant temperature.

Bohr complementarity principle states that quantum systems have conflicting or complementary properties such that the accurate observation of one of these precludes the possibility of observing the other.

Bohr correspondence principle states that for large quantum numbers quantum mechanical and classical predictions must correspond (i.e. agree).

Bohr model is a model of the atom in which the electrons orbit the nucleus in very definite orbits and do not radiate unless they drop from a higher to a lower orbit.

Bohr-Sommerfeld theory is a generalization of the Bohr model and allows calculations for more complicated orbits, including elliptical orbits.

Cathode rays are a beam of electrons emitted from a heated negatively charged electrode.

Cherenkov radiation is also called the "Cherenkov effect". It is the radiation emitted by a charged particle moving through a medium at a speed greater than the speed of light in that medium.

Cloud chamber is a device in which the tracks of charged particles become visible by passing through a supersaturated vapor and producing condensation droplets along the path of the particle.

Complementarity principle See Bohr complementarity principle.

Compton effect is the elastic scattering of a high energy photon by a massive particle such as an electron.

Correspondence principle See Bohr correspondence principle.

Cosmic rays are high energy particles, which exist throughout space.

Electrodynamics is the study of the interaction between electric, magnetic, and mechanical phenomena.

EPR paradox is an apparent paradox described in a paper by Einstein, Rosen, and Podolsky in which they claim to show that the quantum mechanical description of reality is not complete.

Equipartition Theorem states that the average energy of molecules in equilibrium is divided equally among all the possible motions of these molecules.

Exclusion principle See Pauli exclusion principle.

Ether is a hypothetical substance once believed necessary for the propagation of electromagnetic waves. It is also called the "luminiferous ether".

Gamma rays (γ-rays) are very high frequency (higher than X-rays) electromagnetic radiation.

Gravitational red shift is the phenomenon that as light falls in a gravitational field it loses energy and so its frequency shifts to the lower (or red) end of the spectrum.

Habilitation In German universities the rank required to be allowed to teach; this requires a second thesis.

Heisenberg uncertainty principle states that any measurement of a dynamical variable of a quantum system affects the system in such a way that other quantities become uncertain.

Kelvin is the unit of temperature on the absolute temperature scale. The increments are the same as in the Celsius temperature scale, but the zero of temperature is $-273.16°C$. The scale is named after William Thomson, Lord Kelvin.

Lorentz Transformations are a set of equations that allow the comparison of physical observations between two unaccelerated observers moving with uniform speed relative to each other.

Luminiferous ether See ether.

Maxwell's equations are the equations that govern all the phenomena of classical electrodynamics.

Michelson-Morley experiment is the experiment that tried to measure the speed of the earth through the ether.

Minkowski space is the flat four-dimensional space in which time is the fourth dimension. It is useful in considerations that involve special relativity.

Mössbauer effect denotes the emission of a gamma ray from a nucleus bound to a crystal so tightly that the whole crystal, rather than just the nucleus, recoils. This produces radiation of very pure frequency.

Pauli exclusion principle states that no two electrons may exist in exactly the same physical state; thus the quantum numbers describing the states of any two electrons may not be identical.

Photoelectric effect is the phenomenon in which electrons are emitted from the surface of a material when it is irradiated with electromagnetic radiation of sufficiently high frequency.

Photon is the quantum of electromagnetic energy emitted from an atom during radiation; also the massless particle that carries this energy.

Poisson bracket is an anti-symmetric combination of any two dynamical variables that occurs in rather abstract formulations of classical mechanics.

Quaternions are a mathematical generalization beyond the real or complex numbers. They were invented by Sir William Rowan Hamilton and are also called "hypercomplex numbers".

Red shift See gravitational red shift.

Schrödinger cat refers to the cat in a hypothetical experiment devised by Schrödinger to show that quantum mechanics is incomplete.

Space quantization refers to the fact that, in the Old Quantum Theory, the elementary magnets in atomic particles could only point in a few specific directions.

Stark effect is the splitting of its spectral lines when an atom is placed in a strong electric field.

Thermodynamics is the branch of science that deals with the study of energy and the transfer and change of heat energy into other forms of energy.

Uncertainty principle See Heisenberg uncertainty principle.

Uniformitarianism is the hypothesis that geological processes have always proceeded at about the same rate as observed today.

Variational principle is a mathematical statement of the physical fact that, as a system evolves, certain dynamical quantities such as energy change very little from either their maximum or minimum value.

Wilson cloud chamber See cloud chamber.

Zeeman effect is the splitting of an atom's spectral lines when it is placed in a strong magnetic field.

17.2 Chronology of Selected Physics Events

1687: Sir Isaac Newton's *Principia Mathematica Philosophiae Naturalis* is published. The ideas in this book open the door to a rigorous theory of dynamics.

1798: Benjamin Thomson, Count Rumford publishes his results showing that by friction alone unlimited amounts of heat can be extracted from a material. His results destroy the caloric theory of heat.

1842: Julius Rupert Mayer calculates the mechanical equivalent of heat on the basis of experimental data on specific heats.

1843: James Joules measures the mechanical equivalent of heat by various means.

1845; John James Waterston submits his paper on the Equipartition Theorem to the Royal Society. It is not published until 1891.

1846: William Thomson (Lord Kelvin) calculates the age of the earth from its temperature to be only 100 million years. This turns out to be far too low.

1860: James Clerk Maxwell publishes his law of distribution of molecular velocities.

1860: Kirchhoff describes blackbody radiation as a universal phenomenon.

1864: Maxwell modifies Ampère's Law and completes the work on the equations describing all electromagnetic phenomena. They become known as "Maxwell's Equations".

1877: Ludwig Boltzmann publishes his H-Theorem, establishing the connection between entropy and probability.

1885: Johann Jakob Balmer publishes his formula for four lines of the hydrogen spectrum and predicts the wavelengths of further lines.

1887: Albert Abraham Michelson and Edward Williams Morley publish their result, which indicates no detectable velocity for the earth's motion through the ether.

1887: Heinrich Rudolf Hertz publishes the first of a series of papers; the last is published in 1890. These papers establish many of the properties of electromagnetic waves.

1893: Wilhelm Wien publishes his displacement law for blackbody radiation.

1895: Wilhelm Röntgen publishes his paper on X-rays.

1896: Henri Becquerel discovers radioactivity.

1896: Emil Wiechert announces to the Königsberg Physics-Economics Society the discovery of a particle—the electron—with a mass 2,000 to 4,000 times smaller than that of the hydrogen atom. The particle is an electron. He also publishes some of its properties.

1896: Pieter Zeeman discovers the effect named after him, namely that the spectral lines of atoms are split in a magnetic field.

1896: Wilhelm Wien publishes his law for blackbody radiation.

1897: Sir Joseph John Thomson announces the discovery of the electron and some of its properties.

1900: Lord Rayleigh publishes what comes to be known as the Rayleigh-Jeans law for blackbody radiation.

1900: Max Planck derives the blackbody spectrum. This is the birth of quantum physics.

1901: Guglielmo Marconi sends the first transatlantic radio signal.

1902: Rutherford and Soddy discover transmutation of the elements.

1902: Sir Joseph John Thomson proposes the plum-pudding model of the atom.

1904: Hantaro Nagaoka proposes the "Saturnian" model of the atom.

1905: Albert Einstein publishes his Special Theory of Relativity.

1905: Albert Einstein proposes the concept of photons to explain the photoelectric effect. The name "photon" is not coined until several years later by the physical chemist Gilbert Newton Lewis.

1905: Albert Einstein publishes his paper on Brownian motion. His explanation uses kinetic theory and provides convincing evidence of the reality of atoms.

1905: Hermann Nernst announces the Third Law of Thermodynamics.

1907: Albert Einstein uses the equivalence principle to calculate the gravitational red shift.

1908: Hermann Minkowski unifies space and time in Minkowski space.

1908: Heike Kammerlingh-Onnes manages to liquify helium.

1911: Sir Ernest Rutherford discovers that the atom has a tiny nucleus surrounded by a small number of electrons.

1913: Niels Bohr proposes his model of the hydrogen atom.

1913: Henry Gwyn-Jeffreys Moseley shows that the atomic number Z can be easily measured by looking at certain specific lines in the X-ray spectra of the element. His results provide further justification of the Bohr model.

1914: James Franck and Gustav Hertz, unaware of the Bohr theory, show that the electrons in mercury atoms have discrete energies.

1915: Einstein publishes his General Theory of Relativity.

1916: Robert Andrews Millikan verifies the Einstein equation for the photoelectric effect.

1919: Sir Arthur Eddington measures the deflection of light by a gravitational field and verifies Einstein's prediction.

1921: Otto Stern and Walter Gerlach show that the angular momentum of neutral atoms in a magnetic field is quantized. This is "space quantization".

1923: Count Louis-Vincent de Broglie shows the connection between the momentum and wavelength of a particle.

1923: Arthur Holly Compton verifies the particle nature of photons.

1924: Wolfgang Pauli announces what becomes known as the "Pauli Exclusion Principle" and opens the door for later developments in condensed matter physics.

1924: Pauli gives an explanation of the Zeeman effect and predicts two-valuedness of electron states.

1924: Satyendra Nath Bose gives a new derivation of Planck's law, which leads to Bose-Einstein statistics of photons and prediction of a Bose condensate.

1925: G.E. Uhlenbeck and S. Goudsmit postulate the spin of the electron to explain the anomalous Zeeman effect.

1925: Max Born, Werner Heisenberg, and Pascual Jordan complete the formulation of matrix mechanics.

1925: P.A.M. Dirac publishes his version of quantum mechanics.

1926: Erwin Schrödinger publishes his wave equation and immediately applies it to finding the hydrogen spectrum.

1926: Max Born states the probability interpretation of the wave function.

1927: C.J. Davisson and L.H. Germer discover the wave nature of electrons.

1927: G.P. Thomson verifies the wave nature of electrons.

1927: Werner Heisenberg announces the position-momentum uncertainty relation.

1928: Paul Adrien Maurice Dirac publishes his famous relativistic wave equation for the electron.

1928: George Gamow explains that alpha emission is caused by quantum tunneling.

1935: Albert Einstein, Boris Podolsky, and Nathan Rosen publish the paper (known as the EPR Paradox) concerning non-locality in quantum mechanics.

1935: Erwin Schrödinger publishes his paper containing the Schrödinger cat paradox.

Index

QC 7 .C2354 2011

Made in the USA
Lexington, KY
06 January 2013